Marketing of Processed
FRUIT & VEGETABLE

The Authors

   Monalisa Choudhury a Postgraduate from Gauhati University has completed her LL.B and Ph.D. from the same University. She has published a number of Research Papers in many Journals. At present, she is a Lecturer in Commerce Department, Dispur College, Gauhati.

   Nayan Barua, a product of Gauhati University of 1977 completed his LL.B and Ph.D. from Gauhati University. He was a Lecturer of Gauhati Commerce College from 1978 to 1988. He joined the Postgraduate Department of Commerce, Gauhati University in 1988 and is at present Reader in the Department. He has to his credit 30 Research Papers in National and International Journals. He also has two books into his credit.

Marketing of Processed FRUIT & VEGETABLE

Dr. MONALISA CHOUDHURY
Dr. NAYAN BARUA

2015
Daya Publishing House®
A Division of
Astral Interantional Pvt. Ltd.
New Delhi – 110 002

© 2015 Authors

First Impression, 2006

Second Impression, 2012

Reprinted, 2015

ISBN: 978-93-5124-101-0 (International Edition)

Publisher's note:

Published by : **Daya Publishing House®**
A Division of
Astral International Pvt. Ltd.
– ISO 9001:2008 Certified Company –
4760-61/23, Ansari Road, Darya Ganj
New Delhi-110 002
Ph. 011-43549197, 23278134
E-mail: info@astralint.com
Website: www.astralint.com

Laser Typesetting : Classic Computer Services,
Delhi - 110 035

Printed at : Replika Press Pvt. Ltd.

PRINTED IN INDIA

Acknowledgement

I have the privilege to record my profound gratitude to Dr. Nayan Barua, Reader, Department of Commerce, Gauhati University for his valuable guidance and perpetual inspiration. I owe grateful thanks to him. I also express gratitude to my Colleagues at Dispur College for their help, encouragement and untiring support.

I also thank Mr. Dipak Kr. Bhattacharyya, Nitul Sarma, for typing out this work. My special thanks goes to Mr. Dipu Goswami for giving his valuable time and helping me to complete this work. Last but not be least I offer my gratitude and thanks to my parents and my sister who inspired me all the time.

Monalisa Choudhury

Preface

Assam located in the North eastern part of India is predominantly an agricultural state. Its unique geographical and climatic condition has endowed this land with a wide range of horticultural products. The rich and fertile alluvial soil of this region is very good for the growth of fruits and vegetables. Almost every home in Assam both in rural and urban areas has got some kind of vegetable and fruit cultivation. So from this it can be said that Assam is a natural exchequer of fruit and vegetable cultivation which can serve as raw materials for processed food industry.

But inspite of the availability of the raw materials the fruit and vegetable processing industry in Assam is not prosperous and is still in its infant stage. Though some units have made their way to prosperity but such are quite few in number and their development are not organised and steady. It was understood from a preliminary survey of such units that their main problem is the marketing aspect of the finished products which is related to the financial and infrastructural problem.

Although fruit and vegetable processing units are found in the state of Assam yet they are highly unorganised and in many cases they lack viability. Moreover, most of the horticultural products are wasted due to lack of adequate storage facility. Furthermore, due to lack of processing units most of the fruits and vegetables are subjected to free trade within and outside the state.

Taking in view the economic and social viability of this industry, the government of Assam and NEDFI has considered food processing as the thrust area for development.

Though a few paper work has been done on organisational level in this field but such are not sufficient enough to discuss and proceed with the work. This is also on record that this work happens to be a path finder or pioneering work in the field of fruit and vegetable processing in an individual level. The thrust areas of the book are the various problems faced by the fruit and vegetable processing units in Assam especially marketing, their prospects and suggestions for future growth and development. Further the book also attempts to highlight the various government assistance given to such industries. Furthermore, the book also throws light on the agricultural scenario of the state mostly the horticultural produce which serve as raw materials to this industry.

The book is an unique one and has its own significance which throws a tremendous impact on both the economic and social life of Assam. Apart from solving the unemployment problem the growth of this industry brings a fair amount of revenue to the state. Therefore the outcome of this book is expected not only to develop the fruit and vegetable industry but also the standard of living of the masses and thereby enhance the economic and social upliftment of the country.

Monalisa Choudhury
Nayan Barua

Contents

Chapter 1
Introduction

Food, shelter and clothing are the three basic necessities of mankind. Of these, food is the most outstanding one without which no life can ever survive in this world. In the primitive age food was taken raw. But gradually the life style of human being had changed and so also his food habits. Today modern man insists in moulding his food habits to more and more tastier items. Food processing units have grown up with this attitude of mankind to give him with the tastiest and finest food items. The term food processing is quite a wide one consisting of all the edible items. It covers a wide spectrum of products, of which the fruit and vegetable based products cover a unique position. This industry plays a vital role in the effective utilisation of horticultural products. Besides reducing wastage and losses, the fruit and vegetable based units help in raising rural income by generating employment opportunities. The basic features of food processing units are that these units have seasonality in their production cycle. This is because supply of their raw materials are characterized by seasonality. Another feature of these units is that they are engaged in the processing of highly perishable commodities, hence they require greater speed and care in handling, transportation, storage and processing. The third basic feature is that these units face variability in the quantity and quality of their raw materials. This is due to the changes in weather condition and crop/plant diseases. Again these units are very much labour intensive so they generate much employment opportunities especially in the rural areas. And lastly their development saves the scarce foreign exchange presently used in food imports.

Agricultural Background of the North Eastern Region

The North eastern region of India with a mixed terrain of hills and plains intercepted by a large number of small and big rivers, streams is nature's unique gift for production of number of horticultural crops particularly fruits. The land of this region is very rich and fertile and has a wide number of climatic conditions suitable for various types of fruit crops. The plain and valley land of Assam, Tripura, Manipur are well suited for most of the tropical and sub tropical fruit crops like banana, papaya, litchi, guava etc. While the hilly states of Meghalaya, Manipur, Mizoram, Nagaland and Arunachal Pradesh have immense potential for a wide range of sub tropical and temperate fruits like banana, pineapple, citrus, papaya, pear, plum, peach, apple etc.

This region is equally rich in vegetables. There are several kinds of vegetables growing throughout the year. Its congenial soil and climatic conditions makes it ideally suited for extensive cultivation of vegetables too. Maximum areas are presently covered under potatoes while fresh vegetables are mostly confined to areas around cities and towns. Assam which is one of the prominent states of the North Eastern region is mainly an agricultural one. The main occupation of the rural people who constitute more than three-fourth of the total population of the state is agriculture and more than half of Assam's national income comes from agriculture alone. Along with other agricultural crops horticulture bears a bright future in Assam and it has every opportunity to be developed here as valuable processed food products. Assam has an abundant of fruits and vegetables which serve as raw materials for the fruit and vegetable based units of the state.

Food Processing Units of North East and Assam

Since the climate and soil of the North Eastern region is best suited for the agricultural sector there are abundant resources of agricultural products in this region. The outcome of this leads to the emergence of the food processing units in this region. Though there are great potentialities for the establishment of these units yet they are still at their infant stage. There are many reasons for this. Firstly, the marketing problem faced by these units, secondly the infrastructural hurdles and last but not the least is the financial problem faced by these units. Moreover these units also face difficulty in the matters of preserving the raw materials and finished products. Besides severe competition is being faced by these units with the outside products.

Assam, which is one of the most prominent states in the North east, though agriculturally rich yet its food processing units are not developed. This is again due to the aforesaid problems *i.e.*, marketing, infrastructural and financial problems etc. Above these the food processing units of Assam face another problem *i.e.*, the psychological attitude of the local people that the local products are not adequate in comparison to the outside products.

Significance of the Study

Assam is the land of large variety of horticultural crops. Every year a large quantity of fruit and vegetables are wasted due to lack of storage facilities. Therefore for the proper utilisation of such natural products the growth of fruit and vegetable processing units is of outmost necessity. Moreover the fruit and vegetables are sold at a nominal price due to the lack of adequate storage facility which could fetch a fair price if such were purchased by the processing units. Furthermore the growth of such units would solve the unemployment problem to some extent. The growth of such units would also earn a fair amount of revenue to the state. Their development and growth would further facilitate the growth of ancillary industries such as packaging and labelling industry. Therefore the study has got a huge social and economic significance. The study will act as a medium of guidance to the local entrepreneurs in their effort to establish the fruit and vegetable processing units. Further such study will also help the government to come forward and provide the necessary help required by the entrepreneurs for the development of the industry.

Objective of the Study

Though the whole food processing industry face a huge lot of problem including marketing but mostly the marketing problem is seen in the fruit and vegetable industry. The number of fruit and vegetable processing units in the whole of Assam as on 31st March 1992 is recorded to be only 56 according to SSI source, Guwahati

1993, which have gone upto 87 as on 31st March 1998 as per Directorate of Industries, Government of Assam. This is a very negligible number. Therefore it is seen that the fruits and vegetable based units registered with the District Industries Centre (DIC) are in their worst condition mostly due to the marketing problem which is also related with the infrastructural and financial problem. Keeping in view the various problems the following are the objectives of the study:

1. The study objective will cover the various marketing problems directly and indirectly faced by these units. It is proposed to study the gap between demand and supply of the fruit and vegetable based units in the North East and Assam.

2. It is also proposed to study the growth and development of the food processing units in Assam during the five year plans and also the various schemes and assistance provided by the centre and the state during the plans.

3. The study also covers the future prospects of development of the food processing industry in Assam.

4. The crux of the study will cover the entire gamut of the subject ranging from raw materials preservation, transportation, packaging, distribution, training of sales personnel, promotional problems and competitiveness of the products in terms of quality and price etc.

5. The study also proposes to cover the various assistance provided by different government institutions to the various fruit and vegetable units in the North eastern region and Assam.

Methodology

The Methodology of the study is to form core hypotheses and testing of these hypotheses by conducting survey on the various fruit and vegetable based units. Moreover, questionnaire will be prepared and the entrepreneur will be interviewed to collect the relevant data and facts of the present situation of the fruit and vegetable industry in Assam. Besides this, to study the marketing aspect several market analysis will be conducted. A set of questionnaire will also be served upon a cross section of consumers

covering various social groups and income level categories.
Moreover, information relating to various problems especially
marketing will be collected from the concerned agencies, distributors,
retailers and wholesalers along with the various data available
regarding the fruit and vegetable processing units of Assam.
Secondary data will be collected from the various publications,
periodicals, books and information published by the North Eastern
Council, Economic and Statistics Department Govt of Assam, Indian
Institute of Entrepreneurship, Small Industry Service Institute etc.
Further standard statistical tools will also be used as and when
necessary to process the data and the information.

Though the report of the Directorate of Industries, Government
of Assam shows that the number of fruit and vegetable processing
units in Assam goes to 87. Yet from practical survey it was found
that most of the units are closed down and the addresses of a few
could not be traced out. Therefore 50 per cent of the total units are
surveyed the number of which goes to 43. The names of the units
surveyed belonging to different districts runs are shown in
Table 1.1.

Table 1.1

Sl.No.	Name of the Firm	Name of the Entrepreneurs	Address
Kamrup District			
1.	Modern Food Processing Co-operative Ltd.	Nirupama Baruah	Bonda Industrial Complex, Guwahati
2.	Shanti Sadhana Ashram	Hembhai	Shantiban, P.O. Basistha, Guwahati
3.	Padma Food Processing Industrial Society	Syeda Nilufer Rahman	Survey, Ajanta Path Guwahati-38
4.	Sri Ganesh Products	Monomati Barman	Karbi Path, Hatigarh, Guwahati
5.	Progati Enterprise	Nilima Goswami	Industrial Area, Bamuni-maidam
6.	Digo Food Products	Dipti Bhagawati	Surya Kanta Baruah Road, Rukmini Nagar
7.	Dipika Fruit Preservative	Dipika Boro	Ulubari Kachari Basti

Contd...

Table 1.1–Contd...

Sl.No.	Name of the Firm	Name of the Entrepreneurs	Address
8.	Institute of Food Processing-cum Production Centre	Mousami Goswami	Ananda Nagar, Guwahati
9.	Baruah Association	Bhaskar Bharali	VPO Puruni, Guwahati
10.	Bhaiti Agro Foods	Prakesh Agarwal, Binod Agarwal, Anil Agarwal	R.G. Baruah Road, Bhangagarh, Guwahati
11.	Baruah Enterprise	Phatik Baruah	Dakhin Gaon, Kahilipara, Guwahati
12.	Pia Food Products	Nilakhi Choudhury	Murara, Rangia
13.	Ricky Food Products	Jonali Saikia	Srimantapur, Bhangagarh, Guwahati
14.	Swati Food Products	Kumud Dutta	Madhgharia, Noonmati, Guwahati
Sivasagar District			
15.	M/S R.G. Food Products	Mrs Bandana Sarma	Kharagarh Sensowa P.O. Kaloogaon
16.	M/S Tripti Udyog	Smt. Tulumani Gogoi	Bogidol, Bezgoan, P.O. Bogidol
17.	M/S Baruah Enterprise	Smt. Anuradha Baruah	Khatakhunda No. 1, Village P.O. Sapekhati
Golaghat District			
18.	M/S Rushi Products	Pranay Puzari	Rangamati, Chewrigaon
Sonitpur District			
19.	M/S Mridula Food Processing	Mridula Saikia	VPO-Dipota
20.	M/S Satyam Food Preserve	Sukumar Deka	Chandmari, Tezpur
Tinsukia District			
21.	M/S Rochak Foods	Monaj Pariwal	Segumbari, Margherita
22.	M/S Paul Soda	Uttam Paul	Tinsukia
Dibrugarh District			
23.	M/S Champa Pickles	Champa Rami Das	A.T. Road, Chabua
24.	M/S Testi Fruit	Nami Bailung	Latumani, P.O. Tingkhong

Contd...

Table 1.1–Contd...

Sl.No.	Name of the Firm	Name of the Entrepreneurs	Address
25.	M/S Binu's Food Products	Smt. Binu Bhagawati	Sonari Tiniali, P.O. Parbatpur
26.	M/ Allied Food Products	Smt. Sailya Sarma	Santoshi Maa Road, Chring Chapori
27.	Sarala Fruit Products	Rumi Devi	South Amolapatty, Dibrugarh
Karbi Anglong District			
28.	M/S Minas Food Products	Ms. Minara Begum	M. Azad Road, Aminpatty
Nalbari District			
29.	M/S Kiran Udyog	Tapan Das	Tahiaag
30.	M/S Sunny Fruit Preservation Centre	Ms. Suncy	Nalbari Town
Nagaon District			
31.	M/S Mazza Food Production	Md. Makshad Ahmed	VPO, Puranigudam
32.	M/S Elahi Pickles	Md. Ajizul Haque	Singari
33.	M/S Sur Enterprise	Kusum Gogoi	Nagaon
34.	M/S Prasanti Food Products	Jonali Bora	B.M. Road, Amulapatty
35.	M/S Maa Lakhi Udyog	Smt. Nirmali Gayan	Jakhalabandha, Puranigudam
36.	M/S Nivas Food Products	Miss Nivera Begum	Azad Road, Amulapatty
37.	M/S Titly Fruit Products	Chand Sultana	Itachali, Nagaon
38.	M/S Panarama Food Products	Tarulata Tamuly	Nagaon
39.	Juris Food Products	Juri Khound	Jakhalakbandha, Nagaon
40.	J.M. Food Products	Jharna Medhi	Samoguri, Nagaon
41.	Rekha's Food Products	Rekha Devi	Nagaon Town
Darrang District			
42.	M/S Dhanbala Industries	Smt. Dhanbala Devi	Kathpuri, P.O. Menapara
43.	M/S Deepas Food Industry	Smt. Depali Sarmah	Barangabari

Hypotheses

The proposed study covers the problems and the performance of the fruit and vegetable processing industry specially to test the following hypotheses:

1. That the preservation facilities for fruits and vegetables are not adequate to check wastage and decay.

2. That the existing governmental support to the fruit and vegetable based units are not up to the mark.

3. That the local products are not competitive in terms of quality and costs to edge out the products from outside the state.

4. That the consumer behaviour is not friendly and encouraging to local products.

5. That the fruit and vegetable processing industry suffers heavily due to the lack of infrastructural and auxiliary support.

The study of marketing viability of the fruit and vegetable industry is considered crucial in view of the fact that most of them are in the private sector and are unorganised, which creates a gap in the smooth functioning of these units.

However, some of the food based units are also set up under the Khadi and village industries example ghee, honey, rice milling by paddle tooth etc. This units are not commercially viable and are mostly to propagate Gandhian principles only. The study as it has been proposed will fill the gap in the research field relating to the deficiencies in the functioning of these units and pave a way towards the prosperity of these units and thus will create a brighter future to the entrepreneurs, employees and the people in general in this regard.

Chapter 2

Conceptual Framework of Marketing Management

The end of all production is consumption........marketing is the kingpin that sets the revolving of the economy.

Adam Smith

Evolution of Marketing: A Historical Background

Marketing is indeed an ancient art. It has been practised in one form or the other since the beginning of human civilisation. During the pre-industrial revolution the world was characterised by an agriculture cum handicraft economy. Agriculturist and the craftsman were the main producers of this era. Whatever was produced by the agriculturist the surplus of those were disposed off in his immediate neighbourhood after meeting his own requirement. These products were required in the neighbourhood by those who were not engaged in the same activity. There was no elaborate distribution system as the habits and needs of the people and the prevailing technology did not demand such a system. In other words, marketing under those conditions meant production of the basic necessities of life and exchanging these with known consumer groups in the immediate neighbourhood. This represented the barter stage in the evolution of marketing.

The next stage in the evolution of marketing was that of money economy. The change during this stage was limited to the replacement of the barter system by the money economy. The barter system was very inconvenient for the following reasons:

1. It required a double co-incidence of wants. A man with a goat wanting shoes must find a man with shoes wanting goat before any exchange was possible.

2. Under barter small exchanges were very difficult because the items exchanged could not always be subdivided. It was impossible to sell a part of a goat or a part of a shoe.

3. There was no convenient method of keeping savings in the state.

For the above and such other inconveniences the barter economy was replaced by the money economy. Gradually some standard medium of exchange were used which were useful and desirable by all persons in the community. This led to the invention of money. Money can be defined as anything which is generally accepted by the people in exchange of commodities and services or in repayment of debts. Pricing became the chief mechanism of marketing. Economist defines price as the exchange values of a product or service always expressed in money. It is the mechanism or device for translating into quantitative terms (rupees and paisa) the perceived value of the product to the customer at a point of time. From the marketer's point of view, the price also covers the total market offering *i.e.*, the consumer is also purchasing the information through advertising, sales promotion and personal selling and distribution method that has been adopted. The consumer gets these values and also covers their costs. Therefore price can be defined as the money value of a product or service agreed upon in a market transaction. Price is a matter of vital importance to both the seller and the buyer in the market place. In money economy, without prices there cannot be marketing. Price denotes the value of a product. Only when a buyer and a seller agree on price, exchange of goods and services can take place.

The next stage was the stage of Industrial revolution. It originated towards the end of the 18th century. Far reaching changes took place in this stage. The industrial revolution bore the germ of a new business system. The discovery of powerful machines led to the establishment of large industries, textile mills, steel factories, coal mines and so on. It introduced new products, new system of manufacture, new modes of transportation and methods of communication and brought about sweeping changes in the physical and economic environment of man. Mass production became the order of the day. The industrial revolution giving a great deal of disposable surplus income to a large mass of people sustained the mass production and mass distribution unleashed by it. 'The United Nations was the first of the world's societies to move sharply into maturity with the age of high level mass consumption. The former was followed by the Atlantic community, Japan, Australia, New Zealand and several countries in the Soviet Socialist block which also enjoyed relatively high level of mass consumption'.[1]

The mass production and mass distribution brought by the industrial revolution led to the stage of competition. During the earlier stages, the main task of the industrial firms was distribution of whatever they produced. But in the subsequent stages meeting competition became the chief issue. The industrial firms could no longer confine themselves to distribution. The situation demanded a conscious effort to face the competitors, and the firms had to ensure that their products were accepted in preference to those of their competitors. Attempts had been made at different periods in different countries to adopt some of the principles of marketing or establish institutions much similar to the channels of present day distribution. 'Thus according to the new Encyclopaedia Britannica Macropaedica Vol. II (1973) Chain of stores are known to have operated in China several centuries before christ. In 1643 a chain of pharmacies was formed in Japan. AU BON MARCHE in Paris developed from a large speciality store to a departmental store in 1860.'[2]

According to Peter Drucker, "The first man to see marketing clearly as the unique and central function of the business enterprise and the creation of a customer as the specific job of management,

[1]   Fisk Geogre, Marketing System, An Introductory analysis, University of Pennysylvania, Hoper and Row Publishers, 1967 pp. 35-36.

[2]   Banerjee Mrityunjoy, Essentials of Modern Marketing, Oxford & IBH Publishing Co. Pvt. Ltd., New Delhi, 1988 pp. 2.

was Cyrus Mc Cormick who invented the basic tools of modern marketing, market research and market analysis; the concept of market standing, modern pricing policies, the modern service salesman, parts and service supply to the customer and instalment credit. He is truly the father of Business Management and had done all this by 1850."[3]

'Marketing research departments were also opened in the Curtis Publishing Company in 1917. But as Kotler (1980) observed, the task of these departments then was just to develop information that would make the sales department easy to sell.'[4] 'According to Drucker (1978), The revolution of the American economy since 1900 has in large part been a marketing revolution.....fifty years ago the typical attitude of American business towards marketing was, the sales departments will sell whatever the plant produces. Today it is increasingly our job to produce what the market needs.'[5]

Marketing as a scientific process and an organised activity had its origin about the middle of the current century. After the second world war the size and character of markets changed enormously. There was a substantial increase in population. The disposable income of the average family registered an increase. New industrial concerns sprang up rapidly. A great variety of new products and services strengthened the rapidly developing consumer market. At the same time selling of products and services became unusually difficult because of high intensity of competition. It was a situation of abundant choices to the consumers and the consumer began to occupy a place of unique importance. So the consumer got the best return of every penny spent by him and found himself in a position to bargain. The businessman realised that it was not enough if they somehow made one time sale of their products to the consumer. They insisted on the repeated purchase of the products and as such they ensured that the products should be available at a place

---

[3]   Mamoria, C.B., Mamoria, Satish, Marketing Management, Kitab Mahal, 1997, pp. 31.

[4]   Banerjee Mrityunjoy, Essentials of Modern Marketing, Oxford & IBH Publishing Co. Pvt. Ltd., New Delhi, 1988, pp. 2.

[5]   Mamoria, C.B., Mamoria Satish, Marketing Management, Kitab Mahal, 1997, pp. 33.

convenient to the consumer. In addition they had to make available the products at a price advantageous to the consumer. They also ensured that the problems of the consumers were attended promptly. They also made arrangements for replacing the products and providing after sale services when required. Thus this led to the emergence of marketing.

Definition of Marketing

Numerous definitions have been put forward on the term marketing.

A most familiar definition prepared by the definition committee of the American marketing association (Junior of marketing, October 1948) as follows:

'Marketing is the performance of business activities that direct the flow of goods and services from producer to consumer of user.'[6] Individual authorities like Kotler, Clark, Duddy, Hansen have suggested their individual opinions on the definitions of marketing.

The term Marketing is based on conceptual approach to the subject matter which is more target oriented and the performance of the business is judged by the success or failure of fulfilling the target. It is more an activity based definition than of functional nature. However, the definition of Marketing Management is a purely functional approach and views the subject matter in the context of planning, organising, co-ordinating and controlling of activities.

The American marketing association has approved the following definition of marketing management in 1985.

'Marketing management is the process of planning and executing the conception, pricing, promotion and distribution of ideas, goods and services to create exchange that satisfy individual and organisational objectives.'[7]

Philip Kotler, one of the prominent authorities of marketing management has defined it as follows, "Marketing management is the analysis, planning, implementation and control of programmes

---

[6]   Banerjee Mrityunjoy, Essentials of Modern Marketing, Oxford & IBH Publishing Co. Pvt. Ltd., New Delhi, 1988 pp. 4.

[7]   Kotler Philip, Marketing Management, Analysis, Planning, Implementation and Control, Prentice-hall of India Pvt. Ltd., 1992, pp. 11.

designed to bring about desired exchanges with target audiences for the purpose of mutual or personal gain. It relies heavily on the adaptation and co-ordination of product, price, promotion and place of achieving effective response." To make his definition further clear, he lays down the following facts:

1. It is a management process and includes analysis planning, implementation and control activities.

2. It is a purposive activity which aims at bringing desired exchanges which may be of goods or services.

3. It can be practised by either the seller or the buyer, whoever seeks to stimulate the exchange process.

4. It may be carried on for personal or mutual gain.

5. It stresses the adaptation and coordination of several factors that is product price, promotion and place to achieve the effective response.[8]

Marketing in practice, does not refer to any single activity such as selling, advertising or distribution. It is a total function and is concerned with such activities as product planning, product development, product change, pricing, packaging, sales production, advertising and marketing research. Therefore it includes a major part of the activity area of a modern economy. It involves the interaction of several business activities whose ultimate objective is the gratification of customer needs and desires. By satisfying the existing needs of the consumers and creating new needs and wants for better and improved products marketing sets the pattern of consumption and improves the living standards of the people.

Significance and Benefits of Marketing

Marketing occupies an unique place in business. Any organisation that fulfils itself through marketing a product or service is a business. All organisation in which marketing is either absent or incidental is not a business. It is through marketing that the individual needs and wants are satisfied, be it through producing goods or supplying services. Nothing happens in a country until

---

8   Mamoria, C.B., Joshi, R.L., Principles and Practice of Marketing in India, Kitab Mahal, 1997, pp. 63.

somebody sells something. Hence selling aspect becomes the nerve centre for all human activity. Marketing is the kingpin that revolves the economy of a country. The modern concept of marketing recognises its role as a direct contributor to profits as well as sales volume. In today's highly competitive and changing market condition a company must first determine what it can sell, how much it can sell and what approaches must be used to entice the wary customer. No decision regarding plan, production, purchase, budget, design are to be taken until the basic market determinations have been made. Marketing has even greater importance and significance for the society as a whole than for any of the individual beneficiaries of the marketing process.

1. The nation's income is composed of goods and services which money can buy. Any increase in the efficiency of the marketing process which results in lower cost of distribution and lower prices to consumers really brings about an increase in the national income.

2. A reduction in the cost of marketing is a direct benefit to society. The man who makes such a contribution renders a service as important as that of the inventor of a labour saving device or a new manufacturing process.

3. Marketing process brings new varieties, quality and beneficial goods to consumers. It provides a connecting link between production and consumption.

4. Scientific marketing has a stabilizing effect on the price level. As the producers produce what consumers want and consumers have wide choice of products, there is no frequent ups and downs in prices. Scientific marketing is the antithesis of hoarding, profiteering and black marketing.

5. Marketing brings the products to the very door of the consumer. It measures the imbalance in the supply by making available the surpluses to the deficit areas. Marketing provides better provision of transport facilities and storage and thus develops the trade in perishable goods.

6. Patterns of consumption are determined both by the structure of the marketing system which is set up to carry

the flow of goods and services from producers to consumers and by the value added to these goods and services through performance of marketing activities.

7. Marketing is essential for full and near full employment. In order to have continuous production, there must be continuous marketing only then a high level of business activity can be continued.

8. Marketing is advantageous to both, the nation and the society. It changes public opinion and creates new norms of behaviour, new standards of conduct and new ways of life. The tools of marketing can be used to implement the national policy. It also helps in expanding the home market and provide a more secure base for exports.

9. Marketing also helps in the speedy development of underdeveloped and developing nations. It integrates the various economic sectors of the nation such as agriculture and industry. It makes fullest utilization of the existing assets and the productive capacity. It mobilises unknown and untapped economic energy. It also contributes to the development of entrepreneur and managerial class of people.

In sum, marketing tries to find out the right type of production that the firm should manufacture, the right place where it should be made available for use, the right price at which it is to be made available and the right channel through which it is to be brought to the notice of the consumers.

Five Distinct Concepts of Marketing

Exchange Concept

The exchange concept of marketing indicates the exchange of product between the seller and the buyer. The exchange concept covers only the distribution aspect and the price mechanism involved in marketing. But the marketing process is much broader than exchange. The other important aspects such as value satisfaction, concern for the customers, creative selling and integrated action for serving the customer gets completely ignored in the exchange concept of marketing. To view marketing as merely an exchange process would amount to gross undermining of the essence of marketing.

Production Concept

According to the production concept marketing is an attachment to production. The concept emphasizes that marketing can be managed by managing production. The consumer support those products that are produced in great volume at a low unit cost. Though this concept achieves high production efficiency and a substantial reduction in the unit cost of production along with the distribution task making the products widely available yet it fails to attract customers. Customers are motivated by a variety of other considerations in their purchase. Easy availability and low cost are not only the parameters governing the customers buying action. Therefore production concept is not the right marketing philosophy for the enterprise.

The Product Concept

The product concept lays emphasis on high volume of production and low unit cost of production. This concept advocates the product excellence. Organisations supporting this concept believe that the consumers would automatically purchase products of high quality. They spend money, energy and time on research and development and bring in the variety of new products. This concepts neglects the study of the market and the consumer in depth. They concentrate on the product and almost forget the consumer for whom the products are actually meant. Therefore they fail to produce the products as per the demand of the consumers.

The Sales Concept

The sales concept emphasize that a product is not automatically picked by the customers. The company has to consciously promote and push its products. Heavy advertising, high personal selling, large scale promotion, heavy price discount and strong publicity and public relations are the normal tools used by the organisations that rely on this concept.

But according to many, sales concept is a flawless idea. Selling and marketing is often thought to be synonymous, but in reality there is great difference between the two.

The Marketing Concept

The marketing concept was introduced in marketing philosophy only after 1950. This concept lays emphasis on the customer oriented marketing approach and points out that the primary work of a business enterprise is to study needs, desires and values of the potential customers and on the basis of latest and accurate knowledge of market demand, the enterprise must produce and offer the products which will give the desired satisfaction and services to the customers. The marketing concept seeks to believe that marketing starts with the consumer and ends with the satisfaction of those wants. The consumer is placed both at the beginning and end of the business. According to this concept a business can only succeed by supplying to the customer products and services that are properly designed to service their needs.

The essence of marketing concept is that the customer and not the product shall be the centre or the heart of the entire business system. It lays emphasis on the customer oriented marketing process. This concept provides a totally new approach to business. The marketing concept is the best concept as it benefits the organisation, the consumer and the society as a whole. The organisation that adheres to marketing concept keeps itself abreast to the various changes. As a result, it can quickly respond to changes in buyer behaviour and can rectify any drawback in its products. It also gives great importance to planning, research and innovation and its decision are based on reliable data relating to consumer. A firm which does not practise marketing concept get obsolete very soon. Moreover, the firm practising it gets more and more profit because consumer satisfaction guarantees long term financial success. Further, this concept is beneficial to the consumer too. Low price, better quality and ready stocks at convenient locations are some of the benefits enjoyed by the consumer through this concept. The consumer can make cash or credit purchases. He can also purchase on instalment and can even return the material if not satisfied. Therefore the consumer is the greatest beneficiary of the marketing concept. Besides this, the marketing concept also benefits the society. According to this concept only those goods are to be produced which are required by the consumers and therefore it ensures that the country's economic resources are canalised in the right direction. It also acts as a change agents and a value adder in society and therefore improves the standard of living of the people and the pace of economic development of the society. The marketing concept helps in creating good entrepreneurs and managers in society and makes economic planning more meaningful and relevant to the life of the people.

Modern Scenario of Marketing in the Globalised Economy

Marketing is the performance of business that diverts the flow of goods from the producer to the consumer. The marketing process starts even before the goods go into production. It does not end with the sale but continues till the satisfaction of the consumer is obtained. In order to know business one must understand its purpose and the main purpose of business is to create a customer.

'In the rapid process of industrialisation and growing competition, a need for systematic and scientific marketing is fast increasing. Globalisation, a buzzing word in our present age is supposed to accelerate and increase the level of interdependence and competitiveness among nations. It is a change that is transforming the world economy reflecting in widening and intensifying international linkages in trade and finance. It is being driven by a near universal push towards trade and capital market liberalisation, increasing internationalisation of corporate production, distribution strategies and technological change. To an extent globalisation is having the effect of endogenising policy reform in developing countries. The need of any growing economy like India at present is to make economic development realistic and at the same time produce a clear picture of the effects and economic development in a marketing system, a system of integrating wants, needs and purchasing powers of the consumer with capacity and resources of production. The phenomenon of globalisation would suggest that this trend will continue to develop in an irreversible manner over the course of the next millennium in the words of Prof. Bokonga.'[9]

9 Bokonga, Charles Botombele, Globalisation in action faced with the International Challenges of the Global society and the Future of the Individual at the dawn of the 21st Century, 53/54, Law and Society at 9-14, 1996.

'Butros Butros Ghali, on the occasion of 50th anniversary of UN depicted the true picture and effect of globalisation. He stressed that it is an emotive appeal for global solidarity to be at the heart of the international economy which is currently becoming globalised. And acknowledging the name of international society towards the process of change or of globalisation and is at the same time confronted by a true crisis of identity as a result of a loss of its value system or of its traditional mores, pointed out that globalisation increases both opportunities and unfortunately risks.'[10]

In the modern business world competition is growing at a very fast rate. The producers today are trying to put forward their products as best in the market and while doing so, efforts are been made to capture the market and give highest satisfaction to the consumers. For this, the producers have adopted various means of advertising and have also taken recourse to new styles of packaging. Moreover the role of the state has also become more critical both in helping people and firms to grasp the opportunities of the global market place. Further the world trade organisation (WTO), also plays a significant role in making the world into a truly global market place. In this context, the world trade organisation has provided a greater helping hand to the developing and least developed countries in establishing themselves in the world market. The preamble of the agreement establishing the, WTO has the need for positive efforts, designed to ensure that developing countries and especially the least developed among them secure a share in the growth in international trade commensurate with the needs of their economic development. The measures to achieve this objective are substantial reduction of tariff, reduction of other barriers to trade, elimination of discriminatory treatment in international trade, facilitation of concessions of obligation to developing and least developing countries, technical assistance and favourable treatment to them.

In recent years, there has been a substantial expansion in the advertising media available in India. Today, all the major media, the press, the television, the radios, the cinema and the outdoors are used extensively by the advertisers to reach their target consumers. The main purpose of packaging which is considered as the fifth 'P'

---

[10] Bhandari Surendra. World Trade Organisation and Developing Countries, Deep and Deep Publications Pvt. Ltd. New Delhi, 1998, pp. 289.

of marketing is to carry contents safely. However in modern times packaging means much more. Besides, its basic purpose of functionality the key buzz words today are innovation and design. A product must be packaged in such a way that it marks itself out from the rest. Packaging acquires an important dimension for markets in building their brand. As different brands are kept in a limited shelf space of a neighbourhood store or a supermarket, therefore a superior packaging becomes a critical element that attracts a customer choosing one brand over another. Moreover, in today's environment friendly age, packaging with economic friendly material has become essential. Of the various packaging material the current favourite is plastic–PVC, PET, Polyster, Polypropylene in various forms like bottles films, bags etc. New innovations have also been made in the designing and style of packaging. In this era of globalised economy more and more sophisticated ways of packaging has been developed for the world wide marketing of the products. Therefore marketing policies today has been undergone a profound change in order to successfully tackle the global market. At the present time the packaging, labelling, channelising design and style of the products have adopted new processess in order to capture the global market.

Chapter 3

Agricultural Output of North Eastern Region Vis-à-vis Assam

India is the land of many climates and varieties of soils affording a scope for much diversity in agriculture. The geographic location and physical features largely determine the climate of the country.

The lofty Himalayas run along its entire length in the North. To the South of this barrier are the alluvial plains watered by great rivers. In the further South lies the plateau of peninsular India skirted by narrow coastal strips, the Arabian Sea to the West, the Bay of Bengal to the east and the Indian Ocean to the south.

Physiographic Frame of the North East Region

Location

'The North Eastern region of India lies in between the 21.57° and 28.30°E Latitude and 89.46° and 97.30° Longitudes.'[11] The region comprises seven states known as seven sisters namely Assam,

---

[11] Dr. Shukla S.P., Dr. Agarwal A. K., Agriculture in North Eastern Region, National Publishing House, 1986, pp. 3.

Meghalaya, Manipur, Nagaland, Tripura, Mizoram and Arunachal Pradesh. The region is located in the corner of the Indian union and is connected with the rest of India by a narrow strip of land in North Bengal 'The region covers 8 per cent of the total area of the country.'[12]

Boundaries

The North Eastern region has natural frontiers on three sides. The Northern frontiers of the region is guarded by Assam Himalayas from Sankosh river on the west to the entrance of mighty Brahmaputra into Assam. The region is bounded by Bhutan in the west, Tibet and China in the North and East, Myanmar in the South East and Bangladesh in the South.

Natural Division

Geographically the North east region can be divided into four major divisions, the Assam valley, the Assam Himalayas, the Meghalaya hills and the Eastern highlands.

The Assam valley or the Brahmaputra valley is within the griddle formed by the eastern Himalayas, Patkai and Naga hills and the Garo-Khasi and Mikir hills. The Assam Himalayas is deeply dissected by the rivers which flow southwards to join the mighty Brahmaputra. The Meghalaya hills or Shillong plateau is located to the south of the Assam valley. It includes the Garo hills the Khasi and Jayantia hills, the Mikir hills and the North Cachar hills. To the east of Shillong plateau are young mountain ranges running from North to South. These include the Patkai hills, the Naga hills, the Barail hills the highlands of Manipur and Mizo hills.

Climate

The temperature is moderate in the Assam valley whereas the Purbanchal region enjoys a typical monsoon climate ranging from tropical to temperate conditions.Thick clouds pass over the Shillong plateau during summer monsoon period. These clouds reduces the amount of heat received by the ground during day and reduces radiation of heat during night. 'As a result the daily range of

---

[12] Dr. Shukla, S.P., Dr. Agarwal, A. K., Agriculture in North Eastern Region, National Publishing House, 1986, pp. 3.

temperature is less than 6°C in Shillong during the month of July.'[13] 'Main temperature at Guwahati during August the hottest month is 30°C and in January it is 17°C.'[14*]

'The daily temperature in the plains of Brahmaputra and the Barak valley as well as in Tripura and western portion of Mizo hills is above 15°C in January whereas in other parts of the region the temperature is between 10°C to 15°C.'[15] 'The temperature rises from April and in the month of July except the south eastern portion of Mizo hills and around Shillong, the mean temperature ranges from 25°C to 27.5°C.'[16]

'During the month of October daily mean temperature in the hills areas ranges between 20°C to 25°C whereas in the Brahmaputra and the Barak valley Tripura and western portion of the Mizo hills it is above 25°C.'[17]

Rainfall

Rainfall is the single greatest factor governing the cropping pattern and agricultural practices in the region. Because the agricultural production mostly depends upon the distribution pattern of rains during the crop season the rains are of long continuation which commence in the month of March and last till the middle of October. During the month of March and April the fall is very irregular but from May to September it is more steady. Heavy rainfall within few months from June to September results in water logging and floods. This is one of the most important natural factors that influence agriculture in the region.

The region can be divided into three zones on the basis of rainfall intensities. The first zone comprises of north eastern portion of the region upto north of Dibrugarh. The second zone includes the portion

---

[13]   Dr. Shukla S.P., Dr. Agarwal A.K., Agriculture in North Eastern Region, National Publishing House, 1986, pp. 6.

[14]   *ibid.*

*   The current reading of the temperature at present goes to 37°C

[15]   *Ibid.*

[16]   *Ibid.*

[17]   *Ibid.*

on the south of Dibrugarh and the area covering Kohima in Nagaland, Manipur, a part of north Cachar hills, Barak valley and a portion of Mizo hills and the third zone of Brahmaputra valley Meghalaya and south western portion of Mizo hills and Tripura. The intensity of rainfall differs from one zone to the other.

Soil

Soil constitutes the natural medium which supports the growth of plants on earth's surface. The importance of soil for the growth of plants and crops has been recognised by man since time immemorial. The region possesses four main groups of soil from the viewpoint of agriculture. They are:

1. Laterite soil
2. Red loams and Red sandy loams
3. New Alluvial soils and
4. Old Alluvial soil.

Laterites area formed under the conditions of high rainfall with alternating dry wet periods. These soils cover hilly portion of Garo hills, northern portion of Khasi hills and hilly portion of Assam. Red loams and Red sandy loams which are sedimentary formations derived from crystalline metamorphic rocks grow a large variety of crops under rainfed conditions. The whole of Arunachal Pradesh, Mikir hills, Nagaland and Central strip of Meghalaya has Red loamy soil. The red loamy soil also covers the hilly region around north eastern hill area of Assam in the shape of U and southern half beyond Kohima (in Naga Hills) and Tura (in Garo Hills). The new alluvial soil are formed by the floods of river depositing silt near the river banks. This region being loamy and less acidic in nature is fit for growing rabi crops like mustard and pulses. Generally such soils are largely found near the river Brahmaputra.

The old alluvial soil are of old formation and are found in flood free areas and are of more acidic in nature. The soil of greater part of Assam is of this nature.

Land Use Pattern

The pattern of land use of a region at any particular time is determined by the physical, economic and the institutional framework taken together. In other words, the land use pattern depends upon the physical characteristic of the land, the institutional framework, the structure of the other resources *i.e.,* capital, labour etc. available and the location of the region in relation to the other aspects of economic development that is transport, industry and trade. Table 3.1 shows the land use pattern in the North eastern region.

As per the table the state of Arunachal Pradesh has the largest geographical area of 8,734 thousand hectares, followed by Assam with 7,844 thousand hectares, Meghalaya with 2,234 thousand hectares, Manipur with 2,233 thousand hectares, Mizoram with 2108 thousand hectares, Nagaland with 1,658 thousand hectares and lastly Tripura with 1,049 thousand hectares respectively. Assam has the largest area of land utilisation which is 7,852 thousand hectares followed by Arunachal Pradesh, Meghalaya, Manipur, Mizoram, Nagaland and Tripura. The state of Arunachal Pradesh has the maximum area under forest land which comprises of 5,200 thousand hectares and the state of Manipur has the minimum area under forest which is 602 thousand hectares.

Excluding the forest area the land utilisation area can be broadly classified into three divisions:

1. Area not available for cultivation
2. Other cultivable land excluding fallow land
3. Fallow land.

The area not available for cultivation can be further classified into two groups:

1. Area put to non agricultural uses
2. Barren and uncultivable land.

The state of Assam has highest area under non agricultural uses which is 914 thousand hectares, followed by Tripura, Meghalaya, Arunachal Pradesh, Nagaland, Manipur and Mizoram in the descending manner. Similarly in case of Barren and uncultivable land also Assam accounts the first followed by Manipur, Mizoram, Meghalaya and Arunachal Pradesh. The state of Nagaland has no Barren and uncultivable land whereas the data of Barren and uncultivable land of the state of Tripura is not separately available.

Table 3.1: Land Use Classification in North Eastern Region 1989-90 and 1990-91
(Area in 000' Hectares)

State	Geo-graphical Area	Year	Reporting Area for Land Cultivation	Forest Area	Not Available for Cultivation			Permanent Pastures and Other Grazing Lands	Other Cultivable Land Excluding Fallow Land		Current Fallows	Fallow Land	
					Area put to Non Agricultural Uses	Barren & Uncultivable Land	Total (6+7)		Cultivable Waste Land	Others		Fallow Land Other than Current Fallow	Net Sown Area
1	2	3	4	5	6	7	8	9	10	11	12	13	14
Arunachal Pradesh	8374	1989-90	5544	5200	29	48	77	–	–	44	25	49	149
		1990-91	5544	5200	29	48	77	–	–	44	25	49	149
Assam (c)	7844	1989-90	7852	1984	914	1541	2455	184	104	247	88	84	2706
		1990-91	7852	1984	914	1541	2455	184	104	247	88	84	2706
Manipur (j)	2233	1989-90	2211	602	26	1491	1445	(n)	(n)	24	(a)	–	140
		1990-91	2211	602	26	1491	1445	(n)	(n)	24	(a)	–	140
Meghalaya	2243	1989-90	2239(f)	940	84	142	226	–	492	151	60	167	203
		1990-91	2239(f)	939	84	142	226	–	493	153	59	167	202
Mizoram	2108	1989-90	2102	1303	10	201	211	4	74	3	183	259	65
		1990-91	2102	1303	10	201	211	4	74	3	183	259	65

Contd....

Table 3.1–Contd...

1	2	3	4	5	6	7	8	9	10	11	12	13	14
Nagaland	1658	1989-90	1514	863	28	–	28	–	96	124	113	109	181
		1990-91	1532	862	28	–	28	–	96	125	118	110	190
Tripura		1049	1989-90	1049	606	131	(1)	131	(n)	1	39	1	1
270		1990-91	1049	606	131	(1)	131	(n)	1	39	1	1	270

Note: (a) Below 500ha
(c) Relates to the year 1982-83
(f) Excludes area under the municipality and town committee
(j) Ad-hoc estimates
(L) Not available separately, included under cultivable waste
(n) Included under land under misc. tree crops and groves etc.
(a) Adjusted

Source: Basic Statistics of North Eastern Region, 1999, North Eastern Council, Shillong.

The second category of land includes the cultivable land other than fallow land which consist of the pastures and other grazing land, the cultivable waste land and the other cultivable land besides the grazing ground and cultivable waste land. Only the state of Assam and Mizoram has land under pastures and grazing lands which amounts to 184 thousand hectares and 4 thousand hectares respectively. The state of Arunachal Pradesh, Meghalaya and Nagaland has no pastures and grazing lands while the state of Manipur and Tripura has such areas included under miscellaneous tree crops and groves etc. On the other hand in case of cultivable waste land the state of Meghalaya ranks first followed by Assam, Nagaland and Mizoram. The state of Arunachal Pradesh has no area under cultivable waste land while Manipur has no separate area under cultivable waste land but include the land under miscellaneous crops and groves etc. In case of others (which includes the cultivable waste land) Assam ranks first followed by Meghalaya, Nagaland, Arunachal Pradesh, Tripura, Manipur and Mizoram.

The third category includes the fallow land. The dictionary meaning of the word fallow is cultivable land left unsown. Mizoram ranks first in case of both current fallows and other fallow besides current fallow with 183 thousand hectares and 259 thousand hectares respectively. In case of current fallows Nagaland ranks second followed by Assam, Meghalaya, Arunachal Pradesh while in case of fallows other than current fallows Meghalaya ranks second followed by Nagaland, Assam and Arunachal Pradesh. The state of Assam has the highest net sown area of 2706 thousand hectares followed by Tripura 270 thousand hectares, Meghalaya 203 thousand hectares, Nagaland 190 thousand hectares, Arunachal Pradesh 149 thousand hectares, Manipur 140 thousand hectares and Mizoram 65 thousand hectares.

Agricultural Background of the North Eastern Region

The North Eastern region of India is predominantly agricultural. A vast majority of people live in rural areas whose main source of livelihood is agriculture.

Rice is the main food crop grown in the region. Wheat is also grown in Assam, Arunachal Pradesh, Meghalaya and Tripura. A small quantity of sugarcane is also raised particularly in Assam.

Among the plantation crops tea and jute are the most important. More than fifty per cent of the tea produced in the country comes from this region. Moreover jute, cotton, ginger, potato, tobacco, chillies and tapioca are also grown in different parts of the region. Besides these oranges, bananas, pineapples, coconut, apple, peach, plum, pears, litchi, mango, papaya and other citrus fruits cover a large area of the north eastern region. Table 3.2 shows the various fruit crops of the region.

Rice is the main food crop grown in this region. All the states of the North eastern region produces rice which is 5.92 per cent of the total rice produced in India (Figure calculated from the Table 3.2). Further Maize, Sesamum, Potatoes, Rapeseed and Mustard are also grown in all the states of the region. Besides this, wheat, gram pulses, Groundnuts, Sesamum, Sugarcane, Potato and Soyabean are also grown in different parts of the region. As per the Table 3.2, Assam is the only producer of Sugarcane in the North eastern region.

Fruits and Vegetables of the North Eastern Region

The north eastern region of India with mixed terrain of hills and plains intercepted by a large number of small and big rivers and streams is nature's unique gift for production of number of horticultural crops particularly fruits. The land of this region is very rich and fertile and has a wide number of climatic condition suitable for various types of fruit crops. The plain and valley land of Assam, Tripura, Manipur are well suited for most of the tropical and sub tropical fruit crops like Banana, Pineapple, Citrus, Coconut, Papaya etc., while the hill states of Meghalaya, Manipur, Mizoram, Nagaland and Arunachal Pradesh are rich in sub tropical and temperate fruit crops like banana, pineapple, pear and plum, apple etc. Adaptation of various fruit crops in the different latitudes.

Altitudes	Fruit Crops
Upto 800 metres above	Citrus, Pineapple, Bananas, Papaya, Guava, Mango, Litchi, Jackfruit
800–1500 metres	Pear, Plum, peach, apricot
1500 metres and above	Temperate fruits like apple, walnut and almond in specific location of Arunachal Pradesh and peach, plum, apricot in general.

Table 3.2: Food Crops in North Eastern Region Year 1993-94
000' Tonnes

Types	Arunachal Pradesh	Assam	Manipur	Meghalaya	Mizoram	Nagaland	Tripura	Total	All India
Rice	144.0	3361.1	346.6	114.0	96.7	180.0	436.1	4680.5	78972.4
Maize	46.7	11.6	7.8	20.1	14.2	28.0	1.7	130.0	9479.5
Wheat	8.5	100.0	—	6.2	—	1.02	9.5	125.6	59139.3
Small Millets	21.5	4.5	—	2.4	—	10.0	—	38.4	933.3
Gram	—	1.4	—	0.3	—	1.6	0.4	3.7	4903.8
Pulses	5.3	57.0	—	2.5	10.8	10.0	6.5	92.1	13099.6
Tur	—	4.5	—	0.7	2.0	1.0	0.5	8.7	2700.0
Groundnut	—	—	—	—	—	1.2	2.4	3.6	8854.4
Sesamum	0.6	7.2	0.3	0.7	3.8	1.3	1.4	15.3	852.6
Rapeseed and Mustard	20.0	138.0	1.6	4.0	1.3	6.7	7.4	177.7	4871.6
Linseed	—	3.8	—	—	—	0.5	—	4.3	267.9
Sugarcane	—	1700.0	—	—	—	—	—	1700.0	233042.0
Potato	35.5	387.5	23.1	153.2	1.1	23.0	65.5	688.9	16387.9
Soyabean	3.4	—	—	0.9	1.6	5.4	—	11.3	4626.3
Tapioca	—	8.4	—	22.2	7.4	2.9	—	40.9	5021.9

Source: Basic Statistics of North Eastern Region, 1995, North Eastern Council, Shillong.

There is no systematic and accurate estimate of the area and productions of different fruit crops in the North eastern region. The estimate made by the various agencies vary considerably. Table 3.3 shows the fruit crops in the North eastern region. Almost all the states of North eastern region produce pineapple, orange, banana, mango, guava and papaya. While the hills states of Arunachal Pradesh, Mizoram, Nagaland produce peach and plum. Apple is grown mostly in the states of Arunachal Pradesh and Nagaland. Moreover, Jackfruit, Papaya, Litchi, Pears are also grown in different parts of the region.

The north eastern region has a wide cultivation of vegetables too. There are several kinds of vegetables growing throughout the year. The soil and climate of the region is very much suited for cultivation of vegetables. Table 3.4 shows the vegetables and tuber crops of the north eastern region.

Tomato and cabbage are produced by all the states except Tripura. On the other hand Potato is produced by all the states except Manipur. Besides this, except one or two almost all the states has brinjal, onion and cauliflower cultivation to some extent. Moreover, sweet potato squash and colocasia are also cultivated in the region.

Physiographic Frame of the State of Assam

Location and Boundary

'The state of Assam situated in the North east corner of India lies between 24°8' N and 27°56' N latitude and 89°42'E to 96°E longitude'[18]. It is one of the biggest state of the North eastern region comprising of 23 administrative districts of which 21 in the plains and 2 in the hills. The districts in the plains are Dhubri, Kokrajhar, Bongaigaon, Goalpara, Barpeta, Nalbari, Kamrup, Darrang, Sonitpur, Lakhimpur, Dhemaji, Marigaon, Nagaon, Golaghat, Jorhat, Sibsagar, Dibrugarh, Tinsukia, Karimganj, Hailakandi and Cachar. Karbi Anglong and North Cachar hills are the two hill districts.

Boundaries

The state of Assam is connected with the rest of India by a narrow strip of land. It is surrounded by Bhutan and Arunachal

---

[18]   Das, M.M., Peasant Agriculture in Assam, Inter India Publications, New Delhi, pp. 12.

Table 3.3: Fruit Crops of North Eastern Region

Year 1991–92

Area in Hectare | *Production in Metric Tonnes*

Fruit Crops	Assam	Arunachal Pradesh	Manipur	Meghalaya	Mizoram	Nagaland	Tripura	All India
Pineapple	A12129	3363	6450	8502	737	1017	35600	57056
	P177594	12899	5970	74203	4105	2415	30500	768513
Orange	A 4569	4623	1500	7024	5061	—	7639	143475
	P45996	8256	5750	40885	15664	—	36560	938192
Other Citrus	A8169	5206	2120	7024	5374	1511	120160	386929
Lemon/Lime	P60996	8773	7950	40885	17575	1863	41850	2821880
Banana	A39490	1650	3300	4848	2394	1166	3594	383938
	P518128	7818	8500	60684	9880	1656	25000	7790030
Mango	A800	303	180	—	326	46	4972	1077621
	P4680	248	400	—	1258	24	37150	8752134
Guava	A2870	759	500	—	124	248	829	93977
	P380050	1277	1800	—	453	150	2200	1095309
Litchi	A3968	362	—	—	01	45	11870	49277.5
	P11920	389	—	—	02	118	5650	243856

Contd...

Table 3.3.–Contd...

Fruit Crops		Assam	Arunachal Pradesh	Manipur	Meghalaya	Mizoram	Nagaland	Tripura	All India
Papaya	A	4882	554	310	446	136	5	—	45239
	P	750555	9330	1450	3698	552	85	—	805342
Jackfruit	A	—	699	—	—	120	162	6650	—
	P	—	1143	—	—	707	658	17500	—
Apple	A	—	5122	—	—	—	159.7	—	194560.0
	P	—	9330	—	—	—	185.9	—	1147742.9
Pears	A	—	—	—	—	60	—	—	—
	P	—	—	—	—	120	—	—	—
Plum and Peach	A	—	430	—	—	36	32.7	—	—
	P	—	961	—	—	99	591.5	—	—
Others	A	—	—	6915	—	—	—	1361	—
	P	—	—	16920	—	—	—	1710	—

A: Actual

P: Provisional

Source: Basic Statistics of North Eastern Region, 1995, North Eastern Council, Shillong.

Table 3.4: Vegetables and Tuber Crops in the North Eastern Region

Year 1991-92

Fruit/Nuts		Assam	Arunachal Pradesh	Manipur	Meghalaya	Mizoram	Nagaland	Tripura	All India
		Area in Hectare							Production in Metric Tonnes
Tomato	A	13500	241	1100	370	143	16.5	—	289077
	P	312436	667	2350	2220	448	230	—	4244366
Potato	A	61727	4960	—	17639	457	1400	3100	1135075
	P	473250	32470	—	153159	905	15400	56800	18192976
Cabbage	A	25382	2269	950	1000	165	106	—	178353
	P	432026	8960	11500	10200	1829	2214	—	2796431
Sweet Potato	A	9112	—	—	—	—	37	—	—
	P	29010	—	—	—	—	87.5	—	—
Tapioca	A	2153	—	—	—	—	—	—	—
	P	9204	—	—	—	—	—	—	—
Squash	A	—	—	—	—	623	—	—	—
	P	—	—	—	—	21745	—	—	—
Colocasia	A	—	312	—	—	—	—	—	—
	P	—	1636	—	—	—	—	—	—

Contd...

Table 3.4–Contd...

Fruit Crops	Assam	Arunachal Pradesh	Manipur	Meghalaya	Mizoram	Nagaland	Tripura	All India
Brinjal	A19805	557	—	220	510	22	—	—
	P331602	1946	—	1280	1877	325	—	—
Onion	A6172	—	415	220	82	24	—	331736
	P11987	—	4150	1580	62	165	—	4705662
Cauliflower	A20543	984	804	830	83	—	—	202787
	P394628	3579	6405	9960	404	—	—	2998061
Other Vegetables	A3378	2000	—	—	6580	—	—	—
	P12542	2150	—	—	48269	—	—	—

A: Actual

P: Provisional

Source: Basic Statistics of North Eastern Region, 1995, North Eastern Council, Shillong.

Pradesh in the North, Arunachal Pradesh and Nagaland in the East and south east, Nagaland, Manipur, Mizoram and Meghalaya on the south. The state is bounded by Bangladesh in the west and south west and West Bengal in the west. It has international boundaries with two foreign countries Bhutan and Bangladesh.

Natural Divisions

Assam is divided into three natural divisions. They are:

1. The Brahmaputra Valley,
2. The Barak Valley and
3. The Hill Region.

The Brahmaputra Valley

The Brahmaputra valley is an alluvial plain which lies almost east and west in the lower portion but trends to the North east in the upper portion. The plain is formed up of new and old alluvial deposited by the Brahmaputra river and its innumerable tributaries. The valley is also dotted with large number of scattered hillocks which are part of Meghalaya Plateau. The general gradient of the valley is from north east to south west from Sadiya to Guwahati and there from to the west upto Dhubri. Within the valley the north bank slopes southward and the south bank plain slopes northward. The land surface of the Brahmaputra valley may be divided into five zones running parallel to the Brahmaputra river.

1. The Northern foothill zone of the lesser Himalayas
2. The middle plain of the North bank
3. The active flood plain and charlands
4. The middle plain of the south bank
5. The southern foothill zone

The Northern Foothill Zone

The Northern foothill zone consists of the high but narrow Bhabar zone and flat Tarai zone. The alluvial cones at the piedmont of the Himalayas give rise to the formation of highlands or the Bhabar zone composed of unassorted detritus. The Bhabar zone is surrounded southward by a region called Tarai which is a plain of unhealthy damp soil supporting tall grass.

The Middle Plain of North Bank

Between the tarai in the North and flood plain in the south lies a comparatively high and extensive plain which spreads Eastwest parallel to the course of the Brahmaputra river. This plain is gradually tapering towards east and broad towards west. This plain is densely populated with rich rice field. In the eastern part of Darrang and whole of Lakhimpur the plain is extremely narrow as the flood plain approaches the foothill zone. In some places, such high grounds occur near the bank of the Brahmaputra which are favourable sites of tea plantation.

The Active Food Plain and Charlands

On the both banks or the river Brahmaputra lies an extensive and active flood plain which is south of the middle plain of the North Bank. The flood plain is broken in the lower Brahmaputra valley by occasional hillocks and incipient levees, numerous heels and water logged areas. In the North bank the flood plain is widest in the Barpeta and Dhemaji sub divisions. In the south bank it is extensive from Saikhowaghat to Dibrugarh, over Majuli in the Northern Part of Bokakhat inculding Kaziranga National Park and also from the Confluence of the Buridihing to Neemati.

The Middle Plain of South Bank

The middle plain of the south Bank is generally narrow and uneven in its outline. The continuity of the plain is interrupted by the hills of Karbi Anglong. To the west of Nagaon the plain is a narrow lateral strips due to the projection of the Meghalaya plateau. However to the west of Goalpara town the plain becomes somewhat wide because the Garo hills recede southwards with a sharp bend from south Salmara.

The Southern Foothills Zones

The southern foothill zone consists of innumerable high grounds and isolated hillocks scattered with embayments entering into the Tirap and Naga hills. The high grounds composed of laterite soil and covered with either tea gardens or dense forest. In the districts of Nagaon, Kamrup and Goalpara they are composed of monadrocks, erosional plains, beels and swamps which are mostly covered with forest with occasional rice fields.

The Barak Valley

The Barak valley or two Cachar plain lies to the south of the Meghalaya plateau. This plain is created by a gradational and degradational activities of the river barak and its tributaries. The Cachar plain is surrounded by hills on three sides and open towards the west to the plain of Sylhet district of Bangladesh. The Cachar plain is covered with a network of sluggish streams along with a large number of isolated, low hillocks known as tilas. The central part of the plain is built up of the alluvial deposits which make the region a very rich rice belt. Moreover, the high mounds of the interflunces at the Northern, eastern and southern peripheries are dotted with tea plantation.

The Hills

The hill region of Assam consist of Karbi and North Cachar hills. It may be subdivided into three physiographic units—the Karbi hills, the Hamren hills and the North Cachar hills including the high Barail range. Both the Karbi and Hamren hills are geographically parts of the Meghalaya plateau projected into the Brahmaputra valley towards North east up to the proximity of the south Bank of the Brahmaputra. South of the Karbi hills and south east of the Hamren hills lay the hills of Tertiary origin in the North Cachar hills districts. The Barail is the most important range here.

Climate

The climate of Assam is relatively cool and extensively humid with heavy summer rainfall and drought winter. There are four seasons in a year in Assam. They are:

1. Pre-monsoon
2. Monsoon
3. Retreating Monsoon and
4. Dry winter.

The pre-monsoon starts by early March with excessive transpiration and evaporation with dust storms in the Brahmaputra valley. As the thunder storms increases the dust storms subside towards later April. The south west Monsoon starts by the middle of June and continues up to September. The season is characterised mainly by cloudy weather, higher humidity, heavy rainfall, higher temperature and weak variable surface wind. The Monsoon season is the most important period for agricultural activities in Assam. The south west Monsoon starts to retreat by the middle of September which is followed up to the end of November with fair weather and morning fogs. The intensity of rain decreases during this season. The winter season begins at the end of November and continues to the end of February. This seasons is characterized by cool weather with low temperature along with absence of rainfall and morning fogs. This season is suited for the growth of vegetables. The rainless days helps in the harvesting of winter rice but hinder the growth of rabi crops.

Rainfall

Assam experiences abundant spring and Monsoon rainfall which indicates the reason as to why the state is suited for cultivation of such crops as Rice, Jute and Tea. Rainfall is uniform in all the plain districts except southern part of the Nagaon districts which falls under a rain shadow belt. The districts of Dibrugarh, Sibsagar, Lakhimpur, Goalpara in the Brahmaputra valley and Cachar District in the Barak valley receive very heavy rainfall while the districts of Kamrup, Darrang, Nagaon and Karbi Anglong receive comparatively less rainfall. Most of the rainfall occur during the months of June, July and August whereas the months of December, January and February are almost rainless.

Soil

Soil of Assam may be broadly classified into three types alluvial soil, laterite soil and hill soil. Alluvial soil is the most fertile and extensively distributed throughout the Brahmaputra valley. The alluvial soil is divided into two types. The new alluvial soil is found in the middle plain both in the North and south banks of Brahmaputra and on the charlands and active flood plains of the rivers and tributaries. The old alluvial soil is found along the Northern margin of the north bank middle plain. A narrow patch of it also occurs along the southern Margin of south bank middle plain stretching from goalpara to Dhubri. Due to high fertility the alluvial soil is very rich for agricultural purposes, wide varieties of crops such as rice, sugarcane, banana jute, pulses, tobacco, oil seeds and vegetables grow well in alluvial soil. Moreover the areas of soil constitute rice and jute belts of Assam. Tea is also extensively grown in the upper Brahmaputra valley due to the good phosphoric content in the acidic alluvial soil. Laterite soil are of two types:

1. High level laterite, and
2. Low level laterite.

The high level laterite is quite infertile. It is thin and gravelly with little Moisture retain capacity and poor nutritive substances. Only coarse variety of millets and pulses may be grown on such soil. The low level laterite with heavy loams and clays is good for yielding of crops like cotton, rice, sugarcane, banana and tea. In the hills, fruit trees especially citrus fruits are grown abundantly due to the heavy clays and high percentage of organic matter. The hill soils are found in the slopes and ridges of the Karbi Anglong and North Cachar hills. They are dark coloured fertile loams which can produce crops like cotton, rice, maize, millets, potatoes, vegetables and fruits especially orange and pineapple.

Land Use Pattern

As stated earlier the land use pattern of a region depends upon the physical, economic and the institutional framework taken together. Table 3.5 shows the land use pattern in the state of Assam.

The land classification of the various districts of Assam are shown separately. The total geographical area are ascertained both by professional survey and village papers.

The whole of the area are divided into 4 broad categories:

1. Forest
2. Land not available for cultivation
3. Other uncultivated land including fallow land
4. Fallow land.

This district of Karbi Anglong has the largest area under forest with 310 thousand hectares. This district also stands first on account of Barren and uncultivable land covering 593 thousand hectares. The area under non agricultural uses account for 155 thousand hectares in the district of Sonitpur which stands first among all other district in the whole of Assam. The land not available for cultivation excluding permanent pastures and other grazing grounds, Land under Miscellaneous tree crops and groves and cultivable waste land. The Nalbari district ranks first on account of

Table 3.5: Total Area and Classification of Area of Assam 1993-94
(Area in 000'Hectares)

| District | Total Geographical Area according to Survey | Village Pap-ers | Forest | Not Available for Cultivation | | | Other Cultivated Land Excluding Fallow Land | | | | Fallow Land | | | Net Area Sown | Total Crop Area | Area Sown More than Once |
| | | | | Land Put to Non-Agri-cultural Uses | Barre and Uncul-tivated Land Uses | Total | Perm-anent Past-ures and Other Gra-zing Grou-nds | Land Under Misc. Tree Crops and Groves not Inclu-ded in Wet Area | Culti-vable Waste Land | Total | Fallow Land Other than Current Fallows | Curr-ent Fall-ows | Total | | | |
1	2	3	4	5	6	7	8	9	10	11	12	13	14	15	16	17
Dhubri	284	279	41	55	14	69	3	3	4	10	6	6	12	147	225	78
Kokrajhar	313	313	175	18	20	30	4	4	3	11	2	1	3	86	145	59
Bongaigaon	251	250	53	39	37	76	7	3	5	15	3	6	9	97	151	54
Goalpara	182	184	38	22	33	55	8	6	—	10	1	1	2	79	102	23
Barpeta	324	330	87	23	14	37	13	3	2	18	2	2	4	184	323	139
Nalbari	226	208	18	17	7	24	11	2	1	14	1	1	2	150	198	48

Contd...

Table 3.5–Contd...

1	2	3	4	5	6	7	8	9	10	11	12	13	14	15	16	17
Kamrup	435	446	126	68	11	79	21	26	6	53	4	4	8	180	222	42
Darrang	348	348	28	30	36	66	8	17	10	35	7	4	11	208	269	61
Sontipur	532	532	169	155	19	174	12	6	—	18	1	2	3	168	235	67
Lakhimpur	228	235	47	42	38	80	4	2	2	8	2	3	5	95	164	69
Dhemaji	324	324	65	82	50	132	16	17	18	51	2	6	8	68	96	28
Marigaon	170	159	18	21	5	26	10	6	1	17	1	1	2	96	132	36
Nagaon	383	411	92	31	21	52	10	11	5	26	1	1	2	239	384	145
Golaghat	350	354	152	29	10	39	8	14	6	28	5	10	15	120	168	48
Jorhat	285	285	28	82	21	03	4	9	8	21	5	4	9	124	162	38
Sibsagar	267	260	35	40	9	49	7	21	2	30	3	2	85	141	160	19
Dibrugarh	338	339	32	122	10	132	6	26	8	40	5	4	9	126	167	41
Tinsukia	379	379	156	73	40	13	4	4	4	12	1	1	5	96	133	37
Karbianglong	1043	1033	310	(B)	593	593	(B)	(B)	(B)	(B)	(B)	(B)	(B)	130	181	51
N.C. Hills	489	489	89	(B)	372	372	(B)	(B)	(B)	(B)	(B)	(B)	(B)	28	38	10
Karimganj	181	181	50	23	20	43	3	8	2	13	7	7	14	61	106	45
Hailakandi	133	133	60	9	7	16	1	7	—	8	2	1	3	46	62	16
Cachar	379	378	143	41	42	83	3	18	3	24	9	7	16	112	139	27
Assam	7844	7850	2012	1022	1429	2451	158	213	91	462	70	74	144	2781	3962	1181

Source: Statistical Handbook, Assam, 1997, Directorate of Economics and Statistics Government of Assam.

permanent pastures and other grazing ground and land under miscellaneous tree crops and groves with 21 and 26 thousand hectares respectively. On the other hand, Dhemaji district in the upper Assam ranks first on account of cultivable waste land with 18 thousand hectares. The district of Nagaon has the highest sown area in the state with 239 thousand hectares. The district also ranks first for the total cropped area and the area shown more than once.

Agricultural Scenario of Assam

Assam which is one of the prominent states of the North eastern region of India is mainly an agricultural one. Its unique geographical and climatic condition has endowed this land with wide range of agricultural products. The main occupation of the rural people who constitute 90 per cent of the total population of the state is agriculture and more than half of the national income of Assam comes from agriculture alone. 'The state of Assam has 26,95,941 hectares of agricultural land or 34.38 per cent of land are available for production.'[19]

Table 3.6 shows the production of food and non-food crops in Assam.

The main agricultural products of Assam are rice, maize, pulses, rape and mustard, sugarcane, cotton, jute, tobacco, potatoes and several varieties of fruits among which the most prominent ones are: Banana, Pineapple, Orange, Papaya, Sweet Potato, Tapioca and sugarcane. Rice is the staple food of the people of Assam. There are three varieties of rice grown *viz.* autumn rice, winter rice and summer rice named according to the time of harvesting the crop.

The hill district takes the lead in the production of maize. Pulses are grown in all plain districts particularly Kamrup, Nagaon, and Goalpara. Among the oilseeds, rape and mustard are important. Sesamum, linseed, castor are also grown on a small scale. Sugarcane is cultivated in all the plain districts but it is highly concentrated in the Sibsagar District of Upper Assam. Cotton is grown mainly in the Garo and Mikir hills. However the cotton of Assam being short stapled is not much in demand outside the state.

---

[19] Goswami, P.C. Agriculture in Assam. Assam Institute of Development Studies. 1989, pp. 19.

Table 3.6: Production of Food and Non-food Crops Covered by Crop Forecast in Assam (in 000'tonnes)

Sl.No. Crops		1991–92	1992–93	1993–94	1994–95	1995–96	1996–97
1.	Autumn Rice	494	614	587	619	516	520
2.	Winter Rice	2487	2442	2556	2477	2623	2516
3.	Summer Rice	216	243	219	213	251	292
Total Rice		**3197**	**3299**	**3362**	**3309**	**3390**	**3328**
4.	Maize	12	13	12	13	13	13
5.	Wheat	111	79	101	104	95	117
6.	Other Cereals and Small Millets	6	5	5	5	6	5
Total Cereals		**326**	**3396**	**3480**	**3431**	**3504**	**3463**
7.	Gram	2	2	1	1	1	1
8.	Tur	5	4	4	4	4	5
9.	Rabi Pulses	47	45	51	54	51	62
Total Pulses		**54**	**51**	**56**	**59**	**56**	**68**
Total Food Crops		**3380**	**3447**	**3536**	**3490**	**3560**	**3531**
10.	Sesamum	7	7	7	8	8	8
11.	Rape and Mustard	178	138	132	150	143	12
12.	Linseed	4	4	4	5	5	5
13.	Castor	1	1	1	1	1	1
14.	Coconut (a)	94	103	116	118	—	—
Total Oilseeds (Excluding Coconut)		**190**	**150**	**144**	**164**	**157**	**155**
15.	Cotton (b)	1	1	1	1	1	1
16.	Jute (c)	867	1034	676	925	844	803
17.	Mesta (c)	34	30	27	26	27	27
Total Fibres		**902**	**1065**	**704**	**952**	**872**	**831**
18.	Chillies	8	8	8	9	9	10
19.	Turmeric	5	5	6	7	7	7
20.	Arecanut (a)	55	55	54	57	52	—
Total Condiments and Spices		**68**	**68**	**68**	**73**	**16**	—

Contd...

Table 3.6–Contd....

Sl.No.	Crops	1991–92	1992–93	1993–94	1994–95	1995–96	1996–97
21.	Banana	518	538	572	595	565	575
22.	Pineapple	166	184	188	196	195	208
23.	Orange	49	56	58	58	67	69
24.	Papaya	75	83	106	115	107	110
25.	Potato	473	388	507	567	505	579
26.	Sweet Potato	29	29	30	31	31	32
27.	Tapioca	9	8	9	10	12	11
28.	Onion	12	13	15	16	16	17
Total Fruits and Vegetables		**1331**	**1299**	**1485**	**1588**	**1498**	**1280**
29.	Sugarcane (c)	1454	1548	1374	1505	1490	1601
30.	Tobacco	1	1	1	1	1	1
Total Misc. Crops		**1455**	**1549**	**1375**	**1506**	**1491**	—
31.	Mizer	—	—	—	—	4	4
Total Non Food Grains		—	—	—	—	**4022**	**3872**
All Commodities		—	—	—	—	**7582**	**7403**

Note: (a) in terms of million.

(b) 000' of each 170 kg.

(c) 000' of each 180 kg.

(d) in terms of dry cured.

(e) in terms of cane.

Source: Directorate of Economic and Statistics, Assam, Basic Statistics of Assam, 1997.

Assam accounts second in the production of jute in the whole of the Country and produces about one quarter of the total production of the entire country. Tea another important crop is grown mostly in the Upper Brahmaputra valley. Therefore the state of Assam is very rich in agriculture. Inspite of it the state is not self sufficient in food. Much of the agricultural crops are destroyed every year by floods and other natural calamities which cause seasonal scarcity of food.

The Place of Agriculture in the Economy of the North East Vis-à-vis Assam

Agriculture occupies a very prominent position in the economy of the North eastern region. The dependence on agriculture by the majority of population is much more than in other states because of hilly terrain, lack of transportation and communication facilities, absence of industrialisation and other infrastructures in this region. Though small operational holdings, widespread practice of jhum cultivation, occasional floods, natural calamities like drought and hailstorm are the factors that stands in the way of rapid advancement in the agriculture sector yet more than half of the net national income of the region comes from agriculture alone. Wide varieties of food and non food crops are grown in the region. This includes rice, wheat, gram, rapeseed and mustard, sugarcane, potato, ginger, soyabean, arecanut, jute, cotton lint, rubber and the various horticultural crops. There is no other place in India which has beautiful orchards like the orchards of the North East Region particularly of the tribal area. The climate of the region is suited for the various horticultural crops like fruits, vegetables, flowers, tubers, spices etc.

Assam which links the North East to the rest of India is mainly an agricultural state. 'Agriculture is the mainstay of the state since nearly 76 percent of people earn their livelihood through agriculture.'[20] Paddy, jute, wheat, pulses, oilseeds, tapoica, sugarcane, potato turmeric, ginger, coconut, arecanut and vegetables are the main agricultural produce. Less then 5 per cent of the work force of the state are engaged in manufacturing sector. 'The total cropped area in hectares has been estimated at almost 44 per cent of the total geographical land area of the state.'[21] In fact, the economy of the state is heavily dependent on its agriculture as more than 50 percent of the net income is from agriculture to the state exchequer.'[22] Inspite of this, the technology prevalent in the agriculture sector can still be called largely traditional. The fact that the chunk of the substantial area of the state is marshy leaves little scope for bringing new areas under cultivation in Assam.

---

[20] North East times Daily, 9th April, 1994.

[21] *Ibid.*

[22] *Ibid.*

Chapter 4

Profile and Prospects of Food Processing Industry

Definition of Food Processing

The term 'Food Processing' covers all the processed edible items may it be agro processing, meat processing or dairy products. The processed food sector covers a wide spectrum of product which includes Rice mills, Atta chakkis, Supari making units, Bakeries, Oil mills, Noodles making, Fruit and vegetable processing units. Meat processing, Spice grinding, Confectionery, Sea food, Extruded foods and Soft drinks. However there is no prescribed government guideline regarding the definition of the term 'Food Processing.' So far as the administration of different edible products are concerned, earlier the whole spectrum of edible processed products were under the control of Ministry of Agriculture (including food processing) but now different departments are sub-divided which has been entrusted with the control and administration of different products of which food processing is one which comprise of fruit and vegetable products. Likewise, Dairy products are under the control of the Department of Animal Husbandry. Therefore different products are under the control and administration of different departments.

Various Types of Food Processing Units of Assam

Assam is an agricultural state. Its unique geographical and climatic condition has endowed this land with a wide range of raw materials for the food processing industry. The climate of Assam is also suitable for the processing of edible items. But the growth of the units are not evenly distributed throughout the state. In some districts units processing certain items are yet to be started. The Table 4.1 shows the district wise analysis of the various food processing units.

From the table it can be easily understood that the growth of the units are not evenly distributed in all the districts. For example, fruit and vegetable processing which constitute an unique position in the field of food processing is yet to be started in the districts of Cachar, N.C. Hills, Kokrajhar, Barpeta, Bongaigaon, Dhemaji, Hailakandi, Morigaon etc.

Out of the total 3139 food processing units in Assam there are 1310 Rice mills, 495 Atta chakkis, 45 Supari units, 276 Oil mills, 579 Bakeries, 87 Fruit and vegetable units, 23 Noodles making units, 40 Ice making units, 8 Tea packaging units, 21 Bhujiya and Dalmug units, 83 Spice grinding units, 3 Meat processing units, 8 Confectionery units, 126 Modern atta chakkis and 35 Other food processing units.

District Wise Analysis of the Food Processing Units of Assam

Cachar District

The Cachar district situated in the Barak valley region of Assam has a very scanty number of food processing units. The district has 4 Rice mills, 1 Atta chakkis, 4 Bakeries, 3 Ice making units, 2 Bhujiya and Dalmug units, 13 Spice grinding units and 21 Modern Atta chakkis. The total number of food processing units in the districts stands to only 49. Certain items of food processing are to be started in the Cachar district. These are Supari units, Noodles making, Oil mills Fruit and vegetable Tea packaging, Meat processing and Confectionery.

N.C. Hills

The North Cachar Hills district of Assam with its Capital. Halflong has a very small number of food processing units. The total number of food processing units in the district is only 31. Of these, there are 6 Rice mills, 2 Atta chakkis and 23 Bakeries. Other branches of food processing such as Fruit and vegetable processing, Supari, Oil mills, Noodles, Tea packaging, Spice, Meat processing

Table 4.1: District Analysis of Food Processing Units of Assam Since Inception to 31-3-1998

District	Rice	Atta	Supari	Bakery	Oil Mills	Fruit and Vegetables	Noodles	Ice	Tea Pack	Bhujia Dalmug	Spice	Meat	Sweets	Confec-tionary	Others
Cachar	4	1	–	4	–	–	3	–	2	13	–	–	1	21	49
N.C. Hills	6	2	–	23	–	–	–	–	–	–	–	–	–	–	31
Sonitpur	148	18	–	73	27	2	4	–	–	5	–	–	2	36	315
Goalpara	110	35	3	89	21	2	2	–	–	1	–	–	–	1	264
Kamrup	149	39	9	116	48	14	9	–	8	18	–	3	5	21	440
Lakhimpur	99	7	–	29	14	6	1	–	1	1	–	–	3	1	162
Karbi Anglong	35	4	–	7	8	2	–	–	–	–	–	–	–	–	56
Nagaon	288	72	8	22	48	17	3	–	3	7	1	–	2	11	482
Sibsagar	110	33	–	33	16	12	2	–	–	4	–	–	1	2	215
Dibrugarh	63	41	1	30	16	9	6	3	1	6	1	–	–	29	207
Kokrajhar	19	24	–	12	12	2	2	–	–	–	–	–	–	–	71
Dhubri	21	100	2	17	9	–	2	–	4	–	–	1	1	1	162
Darrang	32	17	3	9	8	1	–	–	1	1	–	–	1	1	74

contd...

Table 4.1–Contd...

District	Rice	Atta	Supari	Bakery	Oil Mills	Fruit and Vegetables	Noodles	Ice	Tea Pack	Bhujia Dalmug	Spice	Meat	Confectionary	Others
Karimganj	17	1	–	21	4	13	2	–	–	4	–	1	1	65
Barpeta	54	21	5	26	21	–	2	–	–	4	–	1	–	134
Jorhat	91	47	–	29	19	1	4	–	1	3	–	–	1	201
Nalbari	26	1	1	8	4	1	3	–	–	–	1	–	5	45
Golaghat	6	6	–	7	–	3	–	–	–	5	–	–	–	34
Bongaigaon	3	7	2	6	–	–	1	–	–	1	–	–	4	21
Dhemaji	4	1	2	4	–	–	1	–	–	–	–	–	–	12
Hailakandi	3	3	9	5	–	–	1	–	–	4	–	–	2	27
Morigaon	17	–	–	–	–	–	–	–	–	–	–	2	1	20
Tinsukia	5	15	–	9	1	2	2	5	–	6	–	–	6	52
Total	**1310**	**495**	**45**	**579**	**276**	**87**	**23**	**8**	**21**	**83**	**3**	**8**	**35**	**3139**

Source: Directorate of Industries, Government of Assam.

etc, are yet to be started in the district. The hilly district of Assam has a tremendous supply of horticultural products which may be used as raw materials for the fruit and vegetable units. But there is not a single fruit and vegetable unit in the district at present.

'However, in the year 1973, the Assam Hill Industries Development Corporation had started a fruit and vegetable processing unit under the brand name kaanch (KAANCH) at Jatinga, N.C. Hills. But the unit today have closed down. The production has been totally stopped. The factory shed have today become a haunted one without any human being visiting it.

The reason behind the failure of kaanch is that in the year 1973 when kaanch was started the price of raw materials locally available was very low. But gradually when production gained its momentum and the unit began to flourish the farmers and local cultivators supplying the raw materials raised the prices of raw materials. As a result, the cost of production became very high. So this was one of the root causes for failure of kaanch brand.

Moreover, there was also some sort of mismanagement in the Assam Hill Industries Development Corporation. Thereafter neither the government nor AHIDC has taken any step for revival of kaanch.'[23]

Sonitpur

The Sonitpur district in the middle Assam has a fairly good number of food processing units in comparison to the previously mentioned two districts. The district has a total of 315 food processing units. Out of these, there are mostly rice mills which stands 148 in number. Next in line are 18 Atta chakkis, 73 Bakeries, 27 Oil mills, 2 Fruit and vegetable processing units, 4 Ice making units, 5 Spice grinding units, 2 other food processing units and 36 Modern atta chakkis. The district does not have any Supari, Noodles, Tea packaging, Bhujiya and Dalmug, Meat processing and Confectionery units.

Goalpara

The Goalpara district of Assam has 110 Rice mills, 35 Atta chakkis, 3 Supari units, 89 Bakeries, 21 Oil mills, 2 Fruit and vegetable

---

[23] Directorate of Industries, Government of Assam.

processing units, 2 Ice making units, 1 Spice grinding unit and 1 Modern Atta chakki. The total number of food processing units in the district stands to 264. Other units like Noodles, Tea packaging, Meat processing, Bhujiya and Dalmug, Confectionery are yet to be started in this district.

Kamrup

The Kamrup district of Assam in which Guwahati the capital city of the state is situated has the second highest number of food processing units in the state. The number stands to 440. The district has almost every type of food processing except tea packaging and Meat processing. There are 149 Rice mills, 39 Atta chakkis, 9 Supari units, 116 Bakeries, 48 Oil mills, 14 Fruit and vegetable processing units, 9 Noodles making unit, 1 Ice making unit, 8 Bhujiya and Dalmug units, 18 Spice grinding units, 3 Confectionery, 21 Modern Atta chakki and 5 other types of food processing units.

Lakhimpur

The Lakhimpur district of Assam has 162 food processing units. Out of this, there are 99 Rice mills, 7 Atta chakkis, 29 Bakeries, 14 Oil mills, 6 Fruit and vegetable processing units, 1 Ice making units, 1 Bhujiya and Dalmug unit, 1 Spice grinding unit, 3 other units and 1 Modern atta chakki. Other type of food processing units such as Supari, Noodles, Tea packaging, Meat processing, Confectinery are absent in the district.

Karbi Anglong

Karbi Anglong which is a hill district of Assam has only 56 food processing units, Out of these, there are 35 Rice mills, 4 Atta chakkis, 7 Bakeries, 8 Oil mills, 2 Fruit and vegetable units. Other types of food processing units such as Supari, Noodles, Ice making, Tea packaging, Bhujiya and Dalmug, Spice grinding, Meat processing, Confectionery, Modern Atta chakkis are yet to be started in the district.

Nagaon

The Nagaon district in the middle Assam has the highest number of food processing units in the state. The number stands to 482. Except Noodles, Tea packaging and Confectionery the district witnesses all types of food processing units. There are 288 Rice mills,

72 Atta chakkis, 8 Supari units, 22 Bakeries, 48 Oil mills, 17 Fruit and vegetable processing units, 3 Ice making units, 3 Bhujiya and Dalmug units, 7 Spice grinding units, 1 Meat processing unit, 11 Modern Atta chakkis and 2 other units.

Sibsagar

Sibsagar district has a total number of 215 food processing units. Of these, there are 110 Rice mills, 33 Atta Chakkis, 33 Bakeries, 16 Oil mills, 12 Fruit and vegetable processing units, 2 Noodles unit, 2 Ice making units, 4 Spice grinding units, 2 Modern atta chakkis and 1 other type of food processing unit. The district does not have Supari, Tea packaging and Confectionery units.

Dibrugarh

The Dibrugarh district in the Upper Assam has a total of 207 food processing units. Of these, there are 63 Rice mills, 41 Atta chakkis, 1 Supari unit, 30 Bakeries, 16 Oil mills, 9 Fruit and vegetable processing units, 6 Noodles making units, 1 Ice making unit, 3 Tea packaging units, 1 Bhujiya and Dalmug unit, 6 Spice grinding unit, 1 Meat processing unit and 29 Modern atta chakkis. The district does not have any confectionery unit.

Kokrajhar

The Kokrajhar district of Assam witnesses a total of 71 food processing units. Out of these, there are 19 Rice mills, 24 Atta chakkis, 12 Bakeries, 12 Oil mills, 2 Fruit and vegetable processing units and 2 Noodles making units. Other units like Supari unit, Ice making, Tea packaging, Bhujiya and Dalmug, Spice grinding, Meat processing, Confectionery and Modern atta chakki are totally absent in the district.

Dhubri

The Dhubri district has a total number of 162 food processing units. The district has 21 Rice mills, 100 Atta chakkis, 2 Supari units, 17 Bakeries, 9 Oil mills, 2 Noodles making units, 4 Ice making units, 4 Bhujiya and Dalmug units, 1 Confectionery, 1 Modern atta chakki and 1 other food processing unit. The district does not have any Fruit and vegetable processing, Tea packaging, Spice grinding and Meat processing unit.

Darrang

Darrang district has got 32 Rice mills, 17 Atta chakkis, 3 Supari units, 9 Bakeries, 8 Oil mills, 1 Fruit and vegetable processing unit, 1 Bhujiya and Dalmug unit, 1 Spice grinding unit, 1 Modern atta chakki and 1 other type of food processing unit. The district does not have any noodles making, Ice making, Tea packaging, Meat processing and confectionary units.

Karimganj

The Karimganj district situated in the Barak valley of Assam has a total number of 65 food processing units. There are at present 17 Rice mills, 1 Atta chakki, 21 Bakery units, 4 Oil mills, 13 Fruit and vegetable processing units, 2 Ice making units, 4 Spice grinding units, 1 Confectionery, 1 Modern Atta chakki and 1 Other type of food processing unit. Supari, Noodles, Tea packaging, Bhujiya and Dalmug and Meat processing units are yet to be started in the district.

Barpeta

The Barpeta district of Assam has 134 food processing units of which there are 54 Rice mills, 21 Atta chakkis, 5 Supari units, 26 Bakery units, 21 Oil mills, 2 Ice making units, 4 Spice grinding units and 1 Confectionery unit. Fruit and vegetable processing, Noodles, Tea packaging, Bhujiya and Dalmug, Meat processing, Modern Atta chakki are yet to be started in the district.

Jorhat

The Jorhat district has a total number of 201 food processing units. These units comprise of 91 Rice mills, 47 Atta chakkis, 29 Bakeries, 19 Oil mills, 1 Fruit and vegetable unit, 4 Ice making units, 1 Bhujiya and Dalmug unit, 3 Spice grinding units, 5 other type of food processing units and 1 Modern Atta chakki. The district does not have any Supari, Noodles, Tea packaging, Meat processing and Confectionery units.

Nalbari

The Nalbari district has 45 food processing units which consist of 26 Rice mills, 1 Atta chakki, 1 Supari unit, 8 Bakeries, 4 Oil mills, 1 Fruit and vegetable units, 3 Ice making units, 1 Meat processing units. Noodles, Tea packaging, Bhujiya and Dalmug, Spice Grinding, Confectionery, Modern Atta chakki are totally absent in the district.

Bongaigaon

There are 21 food processing units in the Bongaigaon district of Assam. Of these, there are 3 Rice mills, 7 Atta chakkis, 2 Supari units, 6 Bakeries, 1 Noodles making unit, 1 Ice making unit and 1 Spice grinding unit. There are no oil mills, fruit and vegetable processing, confectionery, modern atta chakki, Tea Packaging, Bhujiya and Dalmug and Meat processing units in the district.

Dhemaji

The Dhemaji district in the Upper Assam has got the lowest number of food processing units in the state. The total number of such units are only 12. Except 4 rice mills, 1 Atta chakki, 2 Supari units, 4 Bakery units and 1 Ice making unit, the rest of the types of food processing units *i.e.* Oil mills, fruit and vegetable processing units, noodles, tea packaging, bhujiya and dalmug, spice grinding, meat processing, confectionery and modern atta chakki are absent in the district. One of the reasons for the lowest number of food processing units in the district is that the district is visited by severe floods every year which disturbs its normal life and also spoils the agriculture which creates difficulty in the setting up of the food processing units. Moreover, the flood also causes serious problems in the transport system of the district. Dhemaji district often remains cut off from the rest of the country during floods.

Hailakandi

The Hailakandi district situated in the Barak valley region of Assam has a very scanty number of food processing units which stands to 27. The district has 3 Rice mills, 3 Atta chakkis, 9 Supari units, 5 Bakeries, 1 Ice making unit, 4 Spice Grinding unit and 2 Other units. Oil mills, Fruit and vegetable processing, Noodles making, Tea packaging, Bhujiya and Dalmug, Meat processing, Confectionery and Modern Atta Chakki are yet to be started in the district.

Morigaon

The Morigaon district also has a very scanty number of food processing units. The number stands to 20. Except 17 rice mills, 2 confectionery and 1 other type of food processing unit, the district does not have any other food processing units. The district does not have any Atta chakki, Supari, Bakery, Oil mills, Fruit and vegetable processing units, Noodles, Ice making, Tea packaging, Bhujiya and Dalmug, Spice grinding, Meat processing and Modern Atta chakki.

Tinsukia

The Tinsukia district in the Upper Assam has 52 food processing units. Of these there are 5 rice mills, 15 Atta Chakkis, 9 Bakeries, 1 Oil mill, 2 Fruit and vegetable processing units, 1 Noodles making unit, 2 Ice making units, 5 Tea packaging units, 6 Spice grinding units and 6 other types of food processing units. The district does not have any Supari, Bhujiya and Dalmug, Meat processing, Confectionery and Modern Atta chakki.

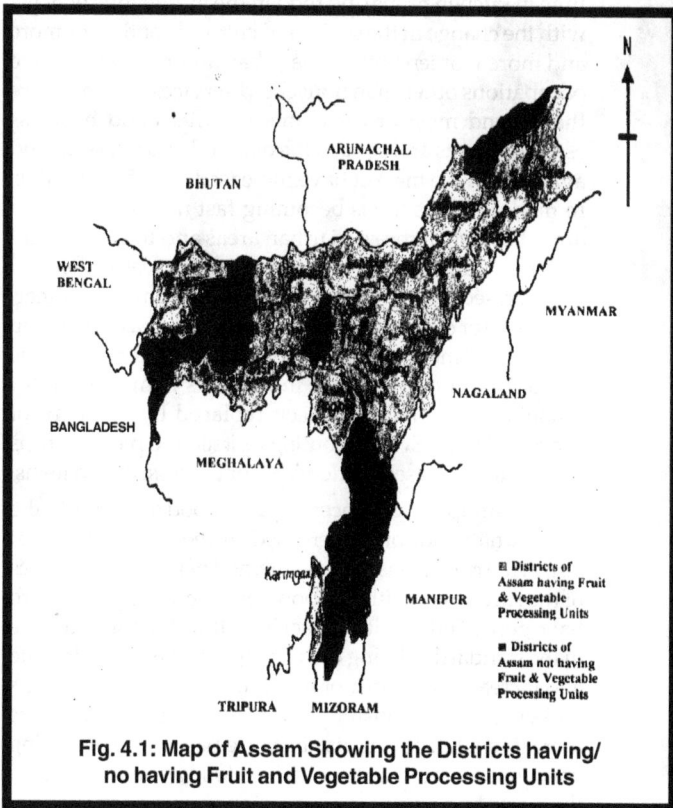

Fig. 4.1: Map of Assam Showing the Districts having/ no having Fruit and Vegetable Processing Units

Prospects of Food Processing Units in Assam

1. With the changing culture and social habits the prospects for setting the food processing units in Assam has increased a lot. Though earlier it was seen that except a few urban areas the consumption of processed food was hardly found in any parts of the state. But today the whole scene have changed. At present with the change in the culture and standard of living of the people both in the urban and rural areas, the consumption and buying of processed food from the market have increased a lot. Though earlier most of the women folk spent most of their time in kitchen preparing food in the indigenous way yet with the change in the social and cultural conditions more and more women both in the urban and rural areas have occupations other than household services and therefore they spend most of their time in offices and business establishments and as such the demand of processed food available in the market have increased a lot. Besides this, in modern times life is becoming fast not only in urban areas but also in the semi urban areas and to some extent in the rural areas too. This also results in increased demand of processed food available in the market. In any seminar, workshop or conference in the semi urban areas the serving of fast food inevitably includes processed food too. The age old habit of preparing indigenous item as tiffin for school children has now been replaced by the items of processed food. Even the outings, picnic and pleasure trips seasonally increases the demand for processed food items.

2. The consumption of processed Baby food available in the market have not only increased in the urban areas but also in the rural areas. The old method of feeding the babies of the rural areas with rice powder ground by pedal tooth have gone. Today with the socio-cultural change and rise in the standard of living of the people and with the fast life and increase in number of working women in both towns and villages, the consumption of instant baby food have increased a lot. This is also because people are becoming very conscious towards nutritious food items available in the market.

3. Assam is an exchequer of the various raw materials required by the food processing industries. Assam is an agricultural state. The rich alluvial soil of the region not only produces rice, wheat but also various horticultural crops which serve as raw materials to the various food processing industry. Since the supply of raw material is in abundance the cost of production is also quite low. Moreover due to the availability of raw materials within the state the cost of transportation of bringing the raw materials to the place of production is also quite cheap. Furthermore, due to the abundance of raw materials the manufacture of wide range of products to be manufactured can be chosen. In this context, mention can be made of the fruit and vegetable processing which is restricted only to a few items. There is a vast potential to exploit minor fruits in the region.

4. The climatic condition of the state is also suitable for food processing.

5. The equipment and machinery required by the food processing industry is not very costly and are available within the country itself. These are mostly procured from the eastern region of the country. Some of the equipments are also available within the state. The procurement of such equipments and machines do not demand a complicated procedure and the cost of transportation is also not very high. Moreover, the machinery and equipments required are easy to maintain and are simple.

6. Food processing units do not demand high technical expertise and also very high qualifications on the part of the entrepreneur. Most of the entrepreneurs of the food processing units are locally trained by the DIC, I.I.E, SISI for a short period and are doing quite well in this sector.

7. Both the Central and the State government have made food processing as one of the thrust areas of development. Large number of incentives and facilities are given by the government in this sector. 'The centre had sanctioned an amount of Rs. 21.41 crores for the various food processing industries in the seven north eastern states during the eight plan of which Assam have received the highest amount of

Rs. 11,84,78,000.'[24] Moreover, in the ninth plan (1997–2002) many incentives and facilities are given to the food processing industries of the state. 'During the first two year of the 9th plan the amount of assistance provided to the whole food processing sector throughout the country stands to Rs. 48 crores which though seem to be a small figure but have generated projects of the value of an estimated Rs. 1100 crore.'[25] 'Furthermore, the government have increased the rate of subsidy upto 35 per cent in case of food processing industry.'[26] Many loans, subsidies and grants are given to the food processing sector of the north-east including Assam. Moreover the government have announced seven years tax holiday for the food processing units in the state of Assam.

8. The food processing industry requires cheap labour which is available within the state itself. The industry does not demand highly skilled labour. Most of the labour of such industry are semi skilled and unskilled who receive training on the job from the entrepreneur. For example in case of fruit and vegetable processing labourers are mostly engaged in doing the cutting work which could be done even by a layman, other works like chopping, grinding, juice extracting being done by the machines.

9. The space required in the food processing unit are relatively smaller than any other units. Most of the units are started by the entrepreneurs in their own household. Though it causes a bit of difficulty but however production could be carried out in a minimum space which is seldom possible in any other industry.

10. It has been found from the field study that the units often fail to supply the products to meet the local demands. Therefore the viability of the units will increase by tapping the local market. Besides this, bright export potentials are also seen in this sector. Here mention can be made of the different food processing products which receive a very good public responses in the fairs and exhibitions outside the state as well as outside the country.

11. The cost of investment in food processing industry is also very low compared to other industry. Moreover, the technology adopted in this industry is also very easy.

12. The large scale expansion of tourism and hotel industry in the entire north eastern region has in turn led to the growth and development of the food processing industry. Today the entire north eastern region has become an important place of tourist resort. Tourist from not only different parts of India but also from abroad come here to see the natural beauty of the region. One of the important point of tourist attraction is the Kaziranga National Park. This in turn has led to the expansion of hotel industry which have led to the increased use of processed food thereby having a tremendous impact on the development of the industry.

Chapter 5

Organisational Problems of Fruits and Vegetable Based Industry in Assam

Unique Characteristics of Fruit and Vegetable Industry

The fruit and vegetable industry fall under the broad category of the agro based industries. The characteristics of the fruit and vegetable industry are as follows:

1. The fruit and vegetable based units have seasonality in their production cycle. This is because the supply of their raw materials are seasonal in nature. This factor compels food processing industries to procure most of their raw material requirements during the harvest season and a little after that so that they can process them during and after the season. The marketing of the processed products is however round the year as their demand is more or less continuous throughout the year.

2. This industry is engaged in the processing of the various fruits and vegetables which are highly perishable in nature. Hence they require greater care in handling, transportation, storage and processing.

3. Location is also an essential feature of the fruit and vegetable industry. Since the raw materials used are highly perishable therefore the location of these units should be in a close proximity to the availability of raw materials. The location should also be near the market to avoid higher transportation cost.

4. Another technical feature of this industry is that unlike other industries, the fruit and vegetable units face variability in the quality and quantity of their raw materials.

5. The manufacturer of the fruit and vegetable based units have no challenging attitude hence there is no cut throat competition among entrepreneurs of agro food products in the domestic markets.

6. The price of the raw materials do not remain steady throughout the year, hence it is difficult for the units to fix the prices of the finished products and contact with the party to supply the products throughout the year at a fixed price.

7. Moreover, the customers are not familiar with frozen foods so the demand for frozen food is very low in India. The customers have also a psychological feeling that frozen foods are not natural but synthetic and contains preservatives.

8. The domestic demand for processed foods is very low because Indians have the habit of eating fresh fruits and moreover the prices of the processed fruits and vegetables are beyond the reach of the common man.

9. This industry also faces many difficulties in developing the domestic market due to lack of infrastructural facilities including cold storage, suitable transport facilities and adequate food testing laboratories.

10. Besides, some companies do not have a strong information network for marketing their products.

11. Another characteristic of this industry is that government play supplementary role in fruit processing by providing facilities to the people at certain agricultural centres at a very nominal cost for consumers. However, such facilities

are very limited and they are confined to only one or two products. Such services rendered at subsidised price create competition for the private enterprises.

Special Features of Fruit and Vegetable Based Units of Assam

Assam which is one of the prominent states of the North eastern region is mainly an agricultural one. Its unique geographical and climatic condition has endowed this land with a wide range of horticultural products. The rich and fertile alluvial soil of this region is good for the growth of vegetables and fruits. Almost every home in Assam wheather in rural or urban areas has got some kind of fruit and vegetable cultivation though in a small sector. For example, Mango, Jackfruit, Arecanut, coconut etc. are found almost in every home. Moreover, vegetables like potatoes, Brinjals, tomatoes etc are also available throughout the state of Assam. So from this it can be said that Assam has a plenty of fruit and vegetables which serve as raw materials for the fruit and vegetable based units of the state. The special characteristics of the fruit and vegetable based units of Assam are as follows:

1. The supply of the raw materials required are seasonal in nature. Therefore the production of these units are subjected to the availability of the raw materials. Moreover, Assam is visited by severe flood havoc every year which damage the crops and create obstacles in the agricultural production of the state.

2. The units are mostly started by the women folk in their own household in an unorganised manner with their own limited capital. Hence the scale of operation of such units are small and could not cover a huge market.

3. Though there is an ample source of raw materials in the state yet the fruit and vegetable based units are very scanty in number. At present there are only 87 registered units all over the state.[27] This is mainly due to marketing problem which is related to the financial and infrastructural problem.

---

[27] Directorate of Industries Reports, 31-3-1998, Government of Assam.

4. Most of the units do not possess the FPO (Fruit products order) licence since the units are mostly started by the entrepreneurs in their own household, the space available for production is very small. But in order to get FPO licence certain minimum space is required (as per fruit products order 1955) which most of the existing units do not fulfil. As a result their products though of good quality is deprived of getting FPO licence. Hence they do not have government preference to market their products within the country and aboard.

5. Moreover, since the raw materials are seasonal in nature therefore the cost of establishment is very high during the off season which increases the maintenance expenses. Furthermore, during the off season when the production is not carried out incessantly the employees have to be paid the same rate of wages which further increases the cost. Besides this, rate of other overhead expenses are to be borne in the same manner as in times of availability of raw materials. This also increases the cost.

6. Instant food is not very popular in the North eastern region including Assam. There are a few urban areas in Assam with a very few middle class people having purchasing power. Furthermore, social life activities in Assam is yet to gain logical momentum with the lifestyle of the people of other region. Though women too in Assam is engaged in different types of services yet most of them have enough leisure time to prepare home made processed food items then to go for those found in the market.

7. There is a common thinking among the people that the fruit and vegetable based items found in the market having chemical preservatives are not good for health. As a result, they prefer the home made processed food.

8. In certain rare instances, there is also a capability of foreign uneatable particles found in the processed foods. This is because the items are often prepared in an unhygienic manner.

9. The processed food items are yet to penetrate into the North eastern market. For example, the mushrooms which are so

popular in other states is yet to gain its popularity in this region.

10. Though Mushroom is regarded as vegetable meat and has a very nutritious food value yet people are suspicious as they think it to be a natural fungus harmful to eat. Moreover, mushroom cultivation is carried in a totally unorganised way in some parts of Kamrup, Nalbari, Nagaon, Karbi Anglong etc. Further, most of units engaged in mushroom cultivation are not registered ones and as such no official data are available as to its volume of production. Besides this, as people are suspicious regarding its consumption therefore the marketing of mushrooms become very difficult. However, the tribal belt of Assam mostly the District of Kokrajhar is a good market for mushroom. Furthermore, Government and the various financial institutions including the banks hesitate to give loans for mushroom cultivation because of its marketing difficulty as its nutritious food value is yet to be popular among the masses.

Production

For the success of any industry production should be carried on in a smooth and uninterrupted way. An uninterrupted production can be only carried on when the purchasing of the raw materials, machines and other equipment required for production is done in an effective manner. Not only this the materials brought are to be stored in an adequate manner to avoid wastage and decay in case the raw materials are perishable in nature. Moreover, the production department has to see whether an efficient quality control process is adopted to test the quality of the finished products. The production activities comprise the following.

Purchase

Every manufacturing process require materials, supplies and services. Before men and machines can start turning out products, materials must be on hand and there must be assurance of a continuing supply to meet production needs and schedules. The quality of materials must be adequate for the intended purpose and suitable for the process and the equipment used. If the material fails on any of these points the results can be costly and expensive production process, inefficient production, inferior products, unheeded delivery promises and unhappy customers. If a company wants to remain competitive and earn satisfactory profits it must procure materials at a lowest cost consistent with quality and service requirements. Efficient inventory management is also a contributory factor for economic and successful production.

In case of food processing industries, raw materials are mostly from agriculture and allied sectors, While processing the raw materials adequate quality, quantity and their availability at an appropriate time and their reasonable cost should be ensured.

The horticultural products such as various fruits and vegetables constitute the main raw materials for the fruit and vegetable processing industry. The supply of these raw materials are characterized by seasonality. Therefore the production is carried on smoothly during this season. Though the production is subjected towards different seasonal fruits and vegetables during different seasons, yet during certain seasons the production frequency curve comes to its lowest level. Moreover, the fruit and vegetable processing industry of Assam depends upon the local markets for the supply of raw materials. The quality, quantity and the price of the raw materials are often unpredictable and fluctuating. Furthermore, Assam is a state which has flood havoc every year. As a result the crops are destroyed by the seasonal floods which is a major setback faced by the agriculturists of the state which hampers the supply of the raw materials. Moreover, unorganised marketing stands on the way of proper coordination between the farmers and the processors.

Storage

Storage is the function of receiving, storing and issuing of materials. In almost all industries raw materials represent a very large investment. It is therefore important that strict orderliness and method are employed to ensure accuracy in preservation and safety at all stages of raw materials movement and custody.

The raw materials of fruit and vegetable are mostly perishable and seasonal in nature. Therefore the raw materials are required to be preserved in cold storage to avoid wastage and decay. But the entrepreneurs of the local fruit and vegetable based units with their limited capital neither could afford to have their own cold storage nor the government has come forward to help them in this regard. Moreover, though most of the units are started by the entrepreneurs at their own residential houses with sufficient space for production yet they are carried on in a unorganised way. This is shown in Table 5.1.

Table 5.1: Showing the Opinion of Entrepreneurs in Percentage Regarding the Space Available for Carrying Production of the Local Fruit and Vegetable Based Products

Types of Queries	Opinion of Entrepreneurs	
	Sufficient	Insufficient
Space available for production	67%	33%

Source: Field study.

The above table which depicts the opinion of the entrepreneurs taken during the survey through questionnaire served upon them shows that 67 per cent of the entrepreneurs reveal that the space available for production is sufficient while 33 per cent of them are of the opinion that the space available are insufficient. Though majority are of the opinion that space available are sufficient yet it is found during the field survey that most of the units are started in an unorganised manner. Though the government have provision for allotment of sheds in industrial areas yet the space of such sheds are hardly sufficient and the facility is not practically enjoyed by many fruit and vegetable processing units of the region. In reality, political clout plays a dominant role for any new entrepreneurs to secure a shed in the industrial centres provided by the government.

Quality Control

When refering to a product the term quality signifies the degree of its excellence. The term quality control means all those activities taken to maintain and improve quality.

It has been realised that inspection alone cannot build quality into a product unless quality has been designed and manufactured into it. Therefore quality awareness must begin at the very conception of a product and continue during the various stages of its development and manufacture and even during its usage to provide feed back from the users which is so essential for quality improvement.

The quality cycle begins and ends with the user. It starts when the users need is analysed to design a product for its fulfilment. The cycle ends with the user because the final proof of good quality comes from the acceptance of the product by the user.

Quality control is an essential criterion for the success of any production unit and in matter of food processing this is most important because any defect in quality of the food products would directly hamper the health of the consumers.

But the fruit and vegetable based units of Assam have not yet paid much interest in the quality control of the products, neither are they very conscious about their quality improvement. Most of the units do not process the FPO (Fruit products order) licence which is regarded quality standard for fruit and vegetable products. This is shown in Table 5.2.

Table 5.2: Opinion of Entrepreneurs in Percentage Regarding Obtainment of FPO Licence by the Local Fruit and Vegetable Processing Units of Assam.

Types of Queries	Opinion of Entrepreneurs	
	Yes	No
Obtainment of FPO licence	45%	55%

Source: Field study.

As per table 5.2 most of the units *i.e.* 55 per cent have not obtained FPO licence due to the strict norms put forward by the ministry of food processing for obtaining such quality standard licence. Only 45 per cent of them have fulfilled the norms and have obtained the licence.

Therefore without FPO licence the products could not obtain ISO 9000 which is an internationally accepted quality standard and as a result the products could not be marketed both within and outside the country. Hence they remain concentrated on a particular locality. Further there is no appropriate laboratory for testing the quality of the products.

Transportation

Transportation plays a vital role both in production and marketing of any products. In case of production when the raw materials are bought from far away places there may arise a problem of deterioration and decay in the absence of cold storage vans. This may also lead to increase in transportation cost.

Moreover, machineries and equipment which are mostly bought from outside the state demand high transportation charges. High transportation expenses are also incurred in buying the packing materials. Furthermore, during the floods which occur every year the transportation of the finished goods and raw materials become very difficult both within and outside the state.

Food Technology

The processing fruit and vegetable products which consist of Jam, Jelly, Chutney, Sauce, Pickles etc cannot be kept for a longer period of time without using adequate preservatives.

Preservation of fruits and vegetables both processed and unprocessed can be broadly classified into three groups.

Physical Methods

Removal of Heat by Refrigeration or Freezing

The freezing preservation of vegetables are done as shown below:

Vegetable

↓

Washing

↓

Cutting

↓

Blanching

↓

Packing in pouches

↓

Freezing

↓

Storing in cold

Freezing Preservation of Vegetables

Fruits are also frozen in a similar way. They are choosen at proper stage of maturity with required brix–acid ratio and washed and prepared in unit packs and frozen in a blast freezer after which they are stored in cold. Fruit pulps can also be frozen in blocks of 25 kg in polythene bags and stored at –18°C.

Addition of Heat

Food products can also be preserved by application of heat as in the case of pasteurization, processing (canning, bottling, aseptic packing).

During pasteurization the pulps/Juices are heated to a temperature of 85°C which kills most of the organisms present. In processing all living organism/spores are destroyed and the condition in the can or bottle do not favour the growth of organism. Fruit Juices are processed at temperature raising from 65°–68°C for 20–25 minutes as higher temperature destroys the flavour. Vegetables which are low in acid and often high in protein contain spore bearing bacteria. So they are processed under pressure in steam.

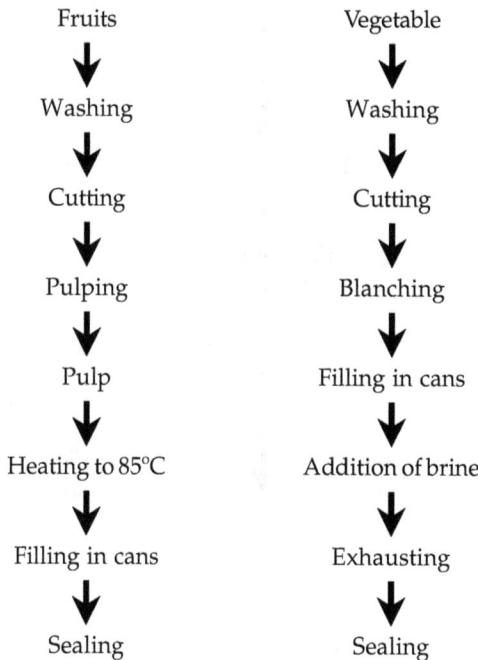

Fruits	Vegetable
↓	↓
Washing	Washing
↓	↓
Cutting	Cutting
↓	↓
Pulping	Blanching
↓	↓
Pulp	Filling in cans
↓	↓
Heating to 85°C	Addition of brine
↓	↓
Filling in cans	Exhausting
↓	↓
Sealing	Sealing

↓   ↓
Processing   Pressure Processing
↓   ↓
Cooling   Cooling
↓   ↓
Storing   Storing

Canning of Fruit Pulp and Vegetables

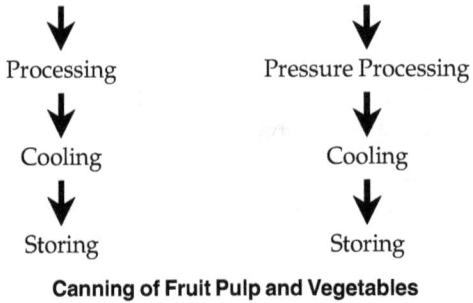

Removal of Water by Dehydration, Drying or Concentration

Preservation by drying depends upon reducing the moisture content to the point at which the concentration of dissolved solids is so high (70 per cent or above) that osmotic pressure will prevent the growth of micro organism. The amount or drying necessary depends upon the composition of food.

Fruits/vegetables
↓
Washing
↓
Peeling/Cutting
↓
Pretreatments
↓
Preservatives (sulphuring, sulphiting)
↓
Drying
↓
Packing
↓
Storing

Drying of Vegetables

Irradiation

In this method the food products is exposed to ultra violet or ionising radiations. This method is commonly used for onion sprout inhibition.

Chemical Methods

Fruits and vegetables can be preserved by addition of preservatives in sufficient concentration to present the growth of micro-organism. The common preservatives used sugar, salt, vinegar, and chemical preservatives like potassium meta bisulphite and sodium benzoate. Sugar used in concentration of 70 per cent or more will preserve most fruits in the form of jellies, jams etc. It acts by osmosis and not as a true micro-organism poison. Salt acts both by osmosis and as micro-organism poison hence it is much more effective than sugar. About 15 per cent salt is sufficient to preserve most products. Acetic acid acts as a micro-organism poison and is much more active than salt. About 2 per cent acetic acid will prevent spoilage. The flow sheets below shows the use of sugar, salt and acetic acid in fruit and vegetable preservation.

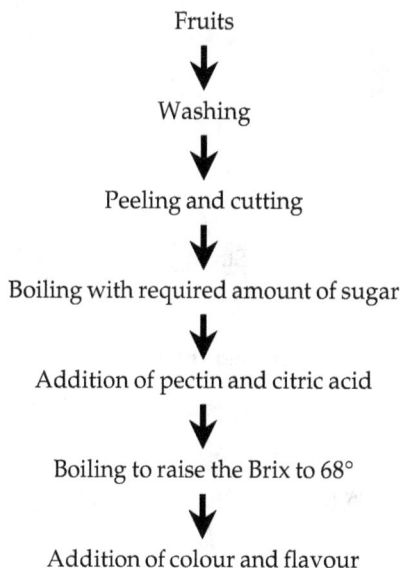

Fruits

↓

Washing

↓

Peeling and cutting

↓

Boiling with required amount of sugar

↓

Addition of pectin and citric acid

↓

Boiling to raise the Brix to 68°

↓

Addition of colour and flavour

↓

Filling in bottles

↓

Storing

Preparation of Jam

Vegetable/raw fruit

↓

Washing and cutting

↓

Addition of salt

↓

Curing

↓

Addition of spices oil

↓

Packing

↓

Storing

Preparation of Pickles

Vegetables

↓

Washing, cutting and pricking

↓

Addition of salt, sugar and acetic acid solution

↓

Filling into jars

↓

Curing

↓

Packing

Preservation of Vegetables Using Acetic Acid

Use of potassium meta bisulphite and sodium Benzoate as preservatives has been widely practised to prevent the spoilages of most of the acid food products. The amount required varies with the pH value (measure of acidity or alkalinity of a solution). The flow chart below represents the preparation of fruit squash and tomato ketchup.

Fruit

↓

Peeling

↓

Pulping

↓

Juice/Pulp

↓

Addition of sugar syrup critic acid, flavour, colour and citric preservatives (potassium meta bisulphite)

↓

Bottling

↓

Storing

Preparation of Squash

Tomatoes

↓

Washing

↓

Juice extraction

↓

Tomato juice

↓

Cooking with sugar, salt, spices

↓

Addition of acetic acid, preservatives (Sodium Benzoate)

↓

Ketchup/Sauce

↓

Hot filling in bottles

↓

Air cooling

↓

Labelling

↓

Storage

Preparation of Ketchup/Sauce

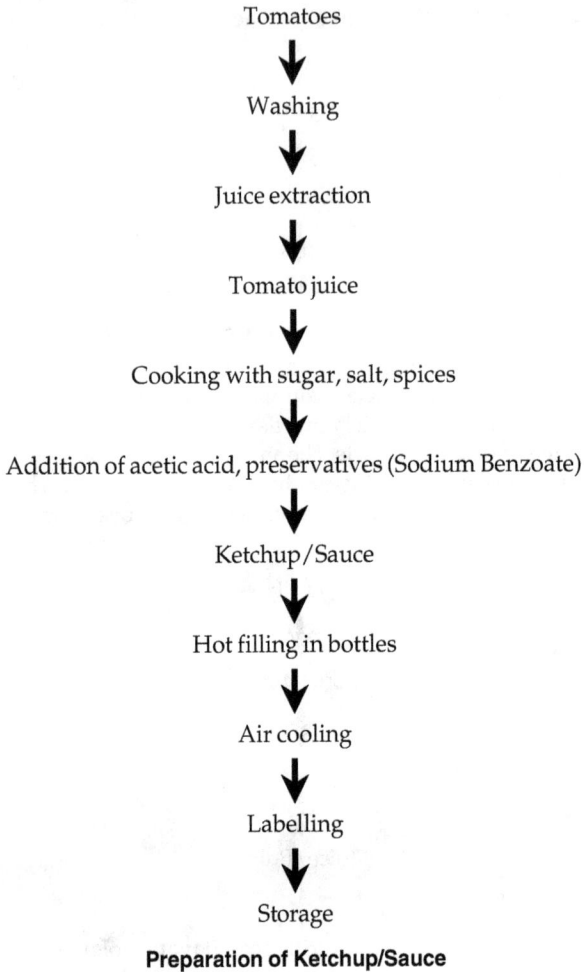

Fermentation

Micro-organism may be used for the preservation of food as well as for their decomposition. Fermentation may be defined as the decomposition of the carbohydrates by micro-organism of enzymes as contrasted with purification.

Alcoholic fermentation by yeast results in the decomposition of the simple hexose sugars into alcohol and carbondioxide. Acetic fermentation resembles alcoholic fermentation and is brought about by vinegar bacteria. Lactic fermentation is extensively used in the preservation of saverkraut, dill ickles, fermented string beans etc.

Requirements for Fruit Products Order Licence

There are five categories of licence under Fruit products order 1955, depending upon the annual production in metric tonnes. The details of the different categories are shown below in the Table 5.3

Table 5.3: Detailed Information of FPO Requirement in Terms of Annual Production in Metric Tonnes

Category	Minimum Manu-facturing Area (Carpet Area) in sq. metre (Flyproof)	Minimum Storage Area (Carpet Area) in sq. metre	Optimum Internal Height of the Manu-facturing Area	Annual Production in Metric Tonnes not to Exceed	Annual Licence Fee in (Rupees)
Large scale	300 sq. mt.	300 sq. mt.	14 ft.	No limit	1500
Small scale (B)	150 sq. mt.	150 sq. mt.	14 ft.	250 M.T.	600
Small scale (A)	100 sq. mt.	100 sq. mt.	14 ft.	100 M.T.	400
Cottage scale	60 sq. mt.	60 sq. mt.	10 ft.	50 M.T.	250
Home scale (B)	25 sq. mt.	25 sq. mt.	10 ft.	10 M.T.	100

Source: Office of the FPO Ministry of Food Processing, Government of India; Six Mile, Khanapara, Guwahati.

Besides satisfying the area and height requirements for the relevant category the building should be:

1. Of permanent structure with RCC, Tin sheets or Asbestos roofing.

2. Well lighted and ventilated.

3. Independent with independent approach from the main road. It shall have no direct communication with residential portion.

4. Location in a clean and hygienic place with surroundings clean and open free from open drains, garbage, cattle sheds etc.

Manufacturing premises should be made fly proof by providing fine mesh on windows external doors, ventilators and openings. Walls should be plastered and white washed, Inner surface of the walls should be made impervious and moisture content upon 1.5 metre height by tiles cement plastered or enamel paint. Floors should be cemented, proper drainage facility should be provided. There should be exit doors (fly proof frame) with auto closing device which open outwards. Moreover there should be provision for free flow of tap water in the manufacturing area. Further the water should be chemically and bacteriologically tested in a recognised laboratory. The water sample for analysis should be drawn by a representative of the laboratory or a local health inspector and mention to this affect is made in the water analysis report.

Finance

Finance is the most essential criterion for the success of any business. It is required at every step of the business may it be for the initial establishment or for modernisation and innovation of the business.

Finance is a major problem faced by the fruit and vegetable based units of Assam. These units are mostly constituted by the entrepreneurs as sole proprietorship concerns with their own capital. Therefore these units always have problems due to lack of finance for which they could not compete with the products from outside the state. Even though various schemes for providing financial assistance to the fruits and vegetable units are introduced by the various government and semi government institutions and some of the units have received such assistance yet the amount of such assistance is very negligible which is not sufficient for proper development of the units. This is shown in Table 5.4.

Table 5.4: Opinion of Entrepreneurs in Percentage Regarding the Financial Assistance Received from Government and Semi Government Institutions

Types of Queries	Opinion of Entrepreneurs	
	Sufficient	Insufficient
Financial assistance received sufficient or not	9%	91%

As per Table 5.4 the conclusion of which is drawn from the field survey carried among the entrepreneurs of the local fruit and vegetable processing units, it is found that 9 per cent of the entrepreneurs surveyed are of the opinion that the financial assistance received from the government and semi government institutions is sufficient while majority of the entrepreneurs *i.e.* 91 per cent are of the opinion that the financial assistance received is insufficient.

Moreover, the rate of interest of such finance is very high and the moratorium period of such finance is very short. Further, financial support requires well throughout project formulation together with quotations of plant from recognised dealers which the entrepreneurs hardly fulfils. Besides this, most of the units are started by the women folk as household units and as part time activity therefore the entrepreneurs of such units cannot give full proof documents relating to business entity and establishment such as registration, assets and other incidental documents fulfilling the norms of financial lending which the financing houses demand for lending finance. Furthermore, since the machineries are not locally produced therefore quotations have to be collected mostly from outside the state which naturally has the problem of cost escalation during the time when the project is finally cleared.

Financial Assistance provided by Various Government and Semi Government Institutions

Both the central and the state government has provided many financial schemes for the development of the fruit and vegetable industry. The government of Assam has increased subsidy upto 35 per cent on investment in all food processing industries.

For providing regular purchase from the farmers and adequate link between the farmer and the processor the government has provided re-imbursement facility in every purchase made by the processor in a given year upto Rs. 10 lakhs for a period of 3 years. Financial assistance is also provided by the Assam financial corporation as term loans. Besides this, Assam Small Industries Development, Corporation, Assam Industrial Development Corporation, Agricultural and Processed Food Products Export Development Authority and other institutions provide financial assistance to the units in different forms. Detailed discussions made in chapter 6.

Moreover, special incentives are provided to the agro and food processing industries which are as follows:

1. Additional state capital investment subsidy of 5 per cent subject to a ceiling of Rs. 5 lakhs for agro and food processing industries.

2. 50 per cent of the cost payable for getting FPO licence/ AGMARK/Trade Mark for the products for food processing industries subject to a ceiling of Rs. 1 lakh.

However, even after such numerous financial assistance and incentive by the Government and the semi Government institutions the food processing units in Assam have not shown much progress. This is because in Assam a common allegation is that the central funds are not properly utilised. The same is witnessed in release of subsidies and incentives owing to the fact that unscrupulous traders along with corrupt officials share a part of it without any purposeful objective. Moreover the support measures adopted by the various Public Sector Undertakings (PSU's) are not fully implemented because many such undertakings are at present in their dying stage.

Personnel

An industrial or commercial organisation comes into existence when a number of persons join hands. These people work to achieve organisational goals. Human resource is of paramount importance for the success of any organisation. It is concerned with the people working in the organisation. Personnel management aims at getting the best result out of the workers. It deals with the obtaining and maintaining of a satisfactory and a satisfied workforce. It is the recruitment, selection, development, utilization of and accommodation to human resources by the organisation.

Labour

Labour is always and everywhere the largest factor of production and labour income always constitute the large part of national income.

Labour represents people employed or capable of being employed in a productive activity. Labour can be divided into two types:

1. Organised labour and
2. Unorganised labour.

The former one is characterized by enforcement of law, security of job, strong bargaining power, better working conditions, uniform wages and so on. The latter is characterized by absence of job security, discriminatory wages, poor working conditions, low bargaining capacity, irregular, disguised and under employment.

The fruit and vegetable based units of Assam enjoys availability in the supply of unskilled labour. There is a dearth of skilled labour required by the fruit and vegetable industry in this region. This is shown in Table 5.5.

Table 5.5: Opinion of Entrepreneurs in Percentage Regarding the Type of Labour available for the Fruit and Vegetable Processing Industry

Types of Queries	Opinion of Entrepreneurs		
	Skilled	Semi Skilled	Unskilled
Type of labour available locally for the fruit and vegetable processing units	18%	37%	45%

Source: Field study.

As per the field survey carried among the entrepreneurs of the local fruit and vegetable processing units, majority of the entrepreneurs are of the opinion that the labourers for such industry are mostly unskilled which is opined by 45 per cent of the entrepreneurs surveyed while 18 per cent of them are of the opinion that labourers are skilled ones. Again 37 per cent of them are of the view that labourers are mostly semi skilled ones. These units which are started by the entrepreneurs in their household are mostly unorganised therefore there is no organised workforce in these units and as such they experience a frequent labour turnover which causes inconvenience in the smooth running of the units. Besides the labourers being mostly unskilled do not have the technical know how of operating the machines. Again since the units are mostly run as household units by the entrepreneurs therefore the production which is carried on a part of the residential house of the entrepreneurs in a minimum space in more or less congested manner is quite unhygienic for the health of the workers.

Wages

Wages is the remuneration paid by the employer to his employee in return for the services rendered by the latter to the former. Wages are usually paid in terms of money soon after the completion of a certain amount of work or a certain period of service like one day, one week, two weeks or one calender month. Thus wages can be defined as a monetary compensation paid by the employer to the worker in return for the contribution he makes for the achievement of the objectives of the organisation.

The wages payable to the workers of the fruit and vegetable based units of Assam differ from unit to unit. There is no uniform standard rate of wages payable to the workers of such units. In some units wages are paid on the basis of fixed monthly salary and in some, payment is made on the basis of daily wages. This is shown in Table 5.6.

Table 5.6: Opinion of Entrepreneurs in Percentage Regarding the Payment of Wages made by them to the Workers of the Fruit and Vegetable Processing Units of Assam

Types of Queries	Opinion of Entrepreneurs		
	Monthly Fixed Salary	Daily Wages	More Work More Pay
Mode of payment of wages by the entrepreneurs of the local fruit and vegetable processing units	64%	27%	9%

Source: Field study.

As per Table 5.6 based on the survey work among the entrepreneurs of the fruit and vegetable processing units of Assam, 64 per cent of the entrepreneurs surveyed prefer to pay wages to workers as fixed monthly salary, 27 per cent of them pay as daily wages and 19 per cent of them pay on the basis of more work and more pay.

The workers of the fruit and vegetable based units of Assam are often exploited and are paid with poor wages. This is because the scale of operation of such units are very small and their magnitude of profit is also very low. As such they cannot pay high wages to the workers. Moreover, no additional benefits are given to the workers along with their wages.

Training

Training is the transfer of defined and measurable knowledge or skills. It involves the development of the workforce to meet the future challenges in the organisation.

So far as the training aspect is concerned the fruit and vegetable based units of Assam have received a very good Government support in this regard. The standard of such trainings as per the entrepreneurs are quite satisfactory. Moreover such trainings have given an added advantage to the entrepreneurs in receiving loans, grants and subsidies from the Government and semi Government institutions. Therefore the entrepreneurs of the local fruit and vegetable based units have not suffered much so far as the training programmes are concerned and most of them have availed the opportunities of such programmes conducted by the Government. However, the entrepreneurs have not yet started any separate training programmes on their own to train out their workers and the local people.

Marketing

Marketing is the sum of all those activities that are related to the free flow of goods from the points of production to the points of consumption. It includes the handling and transportation of goods from the place of production to the place of consumption.

Assam which is the gateway to the North East India has an abundance of horticultural products which serve as raw materials to the fruit and vegetable industry. But the growth of this industry is not proportionate to the availability of the raw materials in this region. This is because of many reasons. Among the various reasons the most vital one is the marketing problem faced by the units. Though the products have a ready market which is of course limited in size yet there are many other marketing problems faced by the units.

Packaging and Labelling

The units face a lot of problems with regards to the packaging and labelling of the products. Containers for packing are not available within the state and are to be purchased from outside which increases the expenditure. Regarding labelling, most of the units prefer ordinary labelling because of its low cost. Moreover the quality of the offset printing of labels done locally are not of good standard. Hence the outside products are more attractively presented so far as packaging and labelling are concerned.

Advertisement

The products of the local fruit and vegetable based units are hardly advertised. As a result, even the local people are not much aware of the products manufactured in the state. It is because of this that the sale of the products are very low compared to the outside products which are rigorously advertised.

Warehousing

The units also face a problem of storing the raw materials and the finished products. Most of the units are started by the entrepreneurs in their own residential houses with minimum space. As a result there is always a problem of storing the raw materials and the finished products. Moreover due to the financial crisis the entrepreneurs could not afford any cold storage facilities. Further the sheds allotted by the government in the various industrial estates are also not sufficient for production and storing purposes.

Market Research

The local fruit and vegetable based units have not shown much interest in conducting any kind of market study for their products. Further no test marketing is done before the launching of the products in the market. However, some units have employed persons for door to door sales of the products who bring certain feedback information regarding the acceptance of the products which to some extent helps in making an analysis of market study. But no services of experts are taken in this regard.

Transportation

Transportation plays a vital role in the marketing of the products. Regarding the marketing of the local fruit and vegetable based products of Assam transportation expenses become very high due to the non availability of the packing materials which are to be purchased from outside the state. Further as the retailers are scattered here and there, therefore, producers have to bear high transportation charges while catering the needs of the individual retailers.

Export Marketing

The products of the local fruit and vegetable units are hardly exported to the other countries. This is because most of the units do not possess the FPO licence. Though a very few negligible number of firms have exported their products on a sample basis yet however the local products are to satisfy the needs of the domestic market fully. Further the entrepreneurs should be rigorously trained to cope up with the various guidelines regarding hygiene conditions which are to be maintained in order to export the products to different countries.

Channel of Distribution

Channel of distribution is the path through which the products pass to reach from the producer to the ultimate consumer.

So far as the local fruit and vegetable based products are concerned the units providing these products have a very small scale of operation and as such these products do not have a very large channel of distribution. The most common channel adopted by these units is the producer → retailer → consumer channel. The entrepreneurs give the products to be sold by the retailers in their own locality. Sometimes it is seen that door-to-door salesman are also employed by the entrepreneurs to sell the products. A few entrepreneurs also have their own retail outlet for selling the products.

Therefore, the fruit and vegetable based units of Assam are besieged with many problems. These units suffer a lot due to the lack of auxiliary support but however they do not have much infrastructural hurdles.

That the fruit and vegetable processing industry suffer heavily due to the lack of infrastructural and auxiliary support.

Table 5.7: Opinion of Entrepreneurs of Fruit and Vegetable Industry in Percentage Showing the Power Supply for Carrying Production in the Local Fruit and Vegetable Industry

Types of Queries	Opinion of Entrepreneurs	
	Sufficient	Insufficient
Power supply for carrying production	73%	27%

Source: Field study

Table 5.8: Opinion of Entrepreneurs of Fruit and Vegetable Industry in Percentage Showing the Transportation System available for the Fruit and Vegetable Industry

Types of Queries	Opinion of Entrepreneurs	
	Yes	No
Any problem with the transportation system	45%	55%

Source: Field study.

Table 5.9: Opinion of Entrepreneurs of the Fruit and Vegetable Industry in Percentage Showing the Cost of Transportation of the Fruit and Vegetable Industry

Types of Queries	Opinion of Entrepreneurs		
	High	Moderate	Low
Opinion about the cost of transportation	45%	37%	18%

Source: Field study.

Table 5.10: Opinion of Entrepreneurs of the Fruit and Vegetable Industry in Percentage Showing the Entrepreneurship Training received by them

Types of Queries	Opinion of Entrepreneurs	
	Yes	No
Whether received any training or not	100	Nil

Source: Field study.

Table 5.11: Opinion of Entrepreneurs of the Fruit and Vegetable Industry in Percentage Showing the Standard of the Entrepreneurship Training

Types of Queries	Opinion of Entrepreneurs	
	Satisfactory	Unsatisfactory
Opinion of the entrepreneurs regarding the standard of the training programme	100%	Nil

Source: Field study.

**Table 5.12: Opinion of Entrepreneurs of the Fruit and Vegetable
Industry in Percentage Showing any Preference
Obtained for Getting (Loans, Grants, Orders)
after receiving the Entrepreneurship Training**

Types of Queries	Opinion of Entrepreneurs	
	Yes	No
Received any kind of assistance (loans, orders, grants) programmes	55%	45%

Source: Field study.

The above tables show the information gathered from the survey work conducted among the entrepreneurs of the fruit and vegetable industry about the infrastructural support received by the units. As per Table 5.7 majority of the entrepreneurs *i.e.* 73 per cent are of the opinion that the power supply to the processing units are sufficient. In case of transport Tables 5.8 and 5.9 shows that although majority of the entrepreneurs are of the opinion that there is no problem in the transport system yet according to them cost of transport is high or moderate. Only few are of the opinion that the cost of transportation is low. Further majority of the entrepreneurs have an optimistic opinion regarding the training programmes. They are of the view that they have received advantages in the form of grants, loans and orders as the result of the training programmes which is shown in the Tables 5.10, 5.11 and 5.12.

Therefore the local fruit and vegetable industry do not suffer much due to the infrastructural support.

**Table 5.13: Opinion of the Entrepreneurs of the Fruit and Vegetable
Industry in Percentage Showing Government Assistance received
by them to promote their sale**

Types of Queries	Opinion of Entrepreneurs	
	Yes	No
Opinion of entrepreneurs as to whether any government or semi government institution have helped them to promote their sale.	27%	73%

Source: Field study.

Table 5.14: Opinion of the Entrepreneurs of the Fruit and Vegetable Industry in Percentage Showing Financial Assistance Received from Any Government or Semi Government Institutions

Types of Queries	Opinion of Entrepreneurs	
	Yes	No
Opinion of entrepreneurs regarding whether any financial assistance received from any government or semi government institutions	64%	36%

Source: Field study.

Table 5.15: Opinion of the Entrepreneurs of the Fruit and Vegetable Industry in Percentage Showing the Financial Assistance Received from the Government

Types of Queries	Opinion of Entrepreneurs	
	Satisfactory	Unsatisfactory
Opinion of entrepreneurs regarding the financial assistance received from the government	9%	91%

Source: Field study.

Table 5.16: Opinion of the Entrepreneurs of the Fruit and Vegetable Industry in Percentage Showing the Credit Facility given to the Retailers

Types of Queries	Opinion of Entrepreneurs	
	Yes	No
Opinion of entrepreneurs as to whether the credit facility is given to the retailers by them	100%	Nil

Source: Field study.

The auxillary support includes the marketing support, the credit facility and the financial assistance provided to the local fruit and vegetable industry.

Table 5.17: Opinion of the Entrepreneurs of the Fruit and Vegetable Industry in Percentage Showing Maximum Credit Period given to the Retailers

Types of Queries	Opinion of Entrepreneurs		
	1 Week	15 Days	1 Month
Opinion of entrepreneurs regarding maximum credit period to the retailers	100%	×	×

Source: Field study.

Table 5.18: Opinion of the Retailers of the Fruit and Vegetable Industry in Percentage Showing as to which of the Products give a Greater Credit Facility

Types of Queries	Opinion of Entrepreneurs	
	Local Products	Outside Products
Opinion of retailers as to greater credit facility	Nil	100%

Source: Field study.

As regards to the marketing support received by the fruit and vegetable industry from the government and semi government institutions as per Table 5.13 majority of the entrepreneurs have received no such support. Only a few *i.e.* 27 per cent of the total entrepreneurs surveyed have received such assistance. Further Tables 5.14 and 5.15 shows that although most of the entrepreneurs have received financial assistance from the government yet the assistance received is not satisfactory. Similarly as per Tables 5.16 and 5.17 though the entrepreneurs give credit facility to the retailers while selling their products yet the periodicity of such credit is very short compared to the products from outside the state. This is due to the shortage of working capital required for the smooth running of any fruit and vegetable industry.

Therefore from the above tables which are drawn from the field study among the entrepreneurs and retailers, a conclusion can be drawn that fruit and vegetable industry suffer heavily due to lack of auxiliary support.

Chapter 6

Role of Government Agencies in the Development of Fruit and Vegetable Industry

Both the central and state government have launched many schemes for the development of the fruit and vegetable industry during the eighth five year plan. 'The central government have sanctioned over Rs. 21.41 crores for various food processing industries including the fruit and vegetable processing industry in the seven north eastern states during the eighth five year plan.'[28]

'Assam has received the highest amount for the development of the food processing industry under the eighth five year plan allocation which is Rs. 11,84,78,000 followed by the other North eastern states.'[29]

Further during the ninth plan too many incentives and facilities were given to the food processing industry both by the central and the state government.

---

[28] North East Time Daily, June 16, 1997.

[29] *Ibid.*

The Government of Assam has also provided many assistance for the development of the fruit and vegetable processing industry of the state. 'The budget for the year 2001–2002 has given its much awaited excise duty exemption to fruit and vegetable processing industry which will make Jams, Jellies, Sauces, Ketchup, Soups and fruit based beverages much cheaper.'[30]

'The state government have increased the subsidy upto 35 per cent on investment in all food processing industries. Moreover, in order to increase the capacity utilisation of fruit and vegetable processing industries by ensuring a regular supply of raw materials and to establish a direct link between the farmer and the processor, the Government has provided incentives in the form of reimbursement upto 5 per cent of the total purchase made by the processor in a given year, limited upto Rs. 10 lakhs per year for a maximum period of 3 years.'[31] But the condition to such reimbursement is that the processing companies will be required to supply high quality seeds, fertilizers pesticides and technology to the contracted farmers along with necessary extension work at a reasonable charge.

Moreover, the government in order to build awareness among consumers about the advantages of processed foods and their quality assurance, have introduced the scheme of generic advertisement. The scheme is a two fold one. Firstly, Generic advertisement and publicity and secondly, market promotion campaign for new product mix and brand name support.

The pattern of assistance are as follows:

1. 'In case of central or state Government Organisation, 50 per cent of the cost of campaign up to Rs. 10 lakhs.

2. In case of NGO's and co-operatives, 50 per cent of the cost of campaign upto Rs. 10 lakhs per annum for a period of 2 years.

3. In case of private industry, assistance is given only for generic advertisement. It consists of 90 per cent of the

---

[30] Baisya, K. Rajat. Implication of Union Budget on Processed Food Industry, Processed Food Industry, Publisher and Printer Salem Akher Taqvi, April 2001, p, 10.

[31] Office of the Directorate of Industries, Government of Assam.

project cost in the first 2 years, 80 per cent of the cost for the next two years and 70 per cent for the last year of the Ninth plan.'[32]

All these assistance are in the form of grants.

Besides this, the government also provide assistance to the fruit and vegetable based units for participating in national and international fairs. All kinds of expenditure are borne by the government in connection with publication of literature, holding seminars, space rentals etc. The quantum of financial assistance depends upon the merits of the proposal. Furthermore, for strengthening the directorate of fruit and vegetable processing including computerisation, compilation of information on different aspects of technology, machinery, packaging etc. the pattern of assistance provided is in the form of 100 per cent grant to both government organisation and private sector.'[33]

'The government also provide assistance for the development of rural entrepreneurship and transfer of technology for processing of food products by utilising locally grown raw materials and providing 'hands on' experience at such production cum training centres; priority being given to SC/ST/OBC and women. The pattern of assistance is in the form of grants upto Rs. 2 lakhs for fixed capital and Rs. 1 lakh for working capital in case of single products line centre. In case of multi product line centre the grant is upto Rs. 750 lakhs for fixed capital and Rs. 2 lakhs for working capital. For training the trainers at recognised institutes such as CFTRI (Central Food Technology Research Institute, Mysore) the grant is upto Rs. 50,000 one time assistance.'[34]

'Besides, the government also provide assistance for setting up mobile fruit and vegetable processing units to take the facility of processing units to the door steps of the farmers. The pattern of assistance of 50 per cent of the project cost (excluding preoperative expenses and margin money for working capital) upto Rs. 40 lakhs in general areas and Rs. 60 lakhs in difficult areas.'[35] Besides, the

---

[32] Office of the Directorate of Industries, Government of Assam.

[33] *Ibid.*

[34] *Ibid.*

government also provide facilities for the establishment of food processing industrial estate/parks with common processing facilities such as analytical and quality control laboratories, cold storage/ modified atmosphere cold storage warehousing facilities, production control facilities etc. 'The pattern of assistance is in the form of grant upto Rs. 4 crores for creation of common facilities.'[36] Moreover, government also give annual awards to the most outstanding units for the achievement in the food processing sector and for augmenting efficiency through a healthy competition.

Besides the direct government assistance, various government and semi government agencies also have schemes for providing various assistance for production and marketing of finished products, imparting training to entrepreneurs of the fruit and vegetable industry, providing financial assistance and also for exporting the finished products outside the country.

North Eastern Regional Agricultural Marketing Corporation Limited (NERAMAC)

The North Eastern Regional Agricultural Marketing Corporation Limited (NERAMAC) was established on 31st March, 1982 by Ministry of Food Processing Industries, Government of India, New Delhi in collaboration with North Eastern Council, Shillong as its promoter. The activities of the corporation include marketing of processed fruit products both in the domestic and international market. The corporation provide regular and systematic marketing facilities to the food processing units of this region to get remunerative prices for their finished products. Moreover, the marketing of ginger from Mizoram and citronella oil from Assam and Nagaland is also being arranged. Further NERAMAC have also arranged export of canned pineapple outside the country. It has also taken up various schemes for export of processed fruit products of the small scale industries of the region thereby ensuring capacity utilisation with commitment of marketing export of agricultural inputs like True potato seeds has also created market for North east origin products in the global market. Furthermore, Neramac has commissioned a

---

[35] Office of the Directorate or Industries. Government of Assam.

[36] *Ibid.*

fruit juice concentration plan in Tripura to market the low volume high valued products of the region. In order to harness the potentiality of cashew cultivation of the region, Neramac has also established a cashew processing unit in Tripura. Further establishment of some more units is on the anvil in other cashew growing areas of the region like Meghalaya.

Apart from marketing of processed fruit products the corporation has undertaken marketing of fresh fruit and vegetables also. Moreover, the corporation is also involved in marketing of agro horticultural inputs, pesticides, seeds, agricultural tools and equipments for agro-horticultural development of the region. The corporation has also taken up marketing of cashew-nut, cereal products, spices and minor forest produces of the North east region.

Assam Small Industries Development Corporation Limited (ASIDC)

Assam Small Industries Development Corporation Limited (ASIDC) was established in the year 1962 as a promotional agency of the government of Assam engaged in development and growth of small industries in the state.

ASIDC has taken up various schemes for providing assistance to the entrepreneurs. These include distribution of raw materials, marketing support, construction and supply of factory sheds, training of workers and supervisory personnels, seed money loan, equity participation, special incentive schemes for scheduled tribe, scheduled caste and women.

ASIDC provides raw materials and various strategic commodities such as paraffin wax, plastic resins, palm fatty acid, non ferrous materials, iron and steel materials, cement, billets etc are made available to SSI units at most reasonable prices. Apart from this ASIDC participates in marketing and tendering of SSI products required by the government departments and agencies. Moreover, factory sheds are provided on nominal rent at various fully equipped industrial areas of the corporation. Such industrial areas are located at Bamunimaidan, Bonda and Jyotinagar around Guwahati and at Sibsagar. Developed lands as per their availability are also provided on lease or sale on hire purchase or on outright basis for construction of factory sheds.

Seed money is given as long term loan to enable the entrepreneurs to meet a part of the margin money called upon by the banks while they extend medium term as well as working capital loan. Besides this, stipendiary training in different trades for personnel to be employed by SSI units are arranged regularly. In case of fruit and vegetable processing units no special type of scheme is adopted by the corporation. The general schemes applicable to all the SSI units are also applied in case of fruit and vegetable processing units.

Assam Financial Corporation (AFC)

The Assam Financial Corporation was set up in the year 1954 under the State Financial Corporation Act 1951 with equity contribution from the government. The aim of the corporation is to provide financial assistance to different production units by way of various schemes.

The area of operation of the Assam Financial Corporation includes four states which are Assam, Meghalaya, Manipur, Tripura. However, the share capital has been contributed only by the Assam, Manipur and Tripura government. The corporation has three regional offices at Tinsukia, Tezpur and Bongaigaon. It has branches at Agartala, Imphal, Shillong, Jorhat, Nagaon, Silchar, Dibrugarh, Sibsagar, North Lakhimpur, Nalbari and Dhubri.

The various schemes of the corporation are the transport loan schemes, Composite loan schemes, Schemes for technocrats and medical Graduates, Schemes for Scheduled caste/scheduled tribe entrepreneurs, Schemes for hospital and Nursing home, Schemes for printing press, small loan, Special capital assistance, Equipment finance, modernisation and expansion scheme.

Different schemes have different terms and conditions along with different rate of interest. The above mentioned schemes were in existence prior to 1993. But from 1994 onwards due to financial crisis, lack of recovery of the loan components and withdrawal of share capital contribution by the government the Assam Financial Corporation have suffered a lot. Therefore from 1994 onwards the various schemes of the corporation were abolished. At present, the corporation have only one master scheme for financing all the production units. Moreover, they strictly observe the viability of the unit before extending any kind of financial assistance to it. The various aspect which the corporation observes before granting any loan are the financial viability of the unit, the marketing aspect, quality production and above all the background of the promoter which includes his/her qualifications, sense of business interest etc.

At present the terms and conditions of granting loans to the production units are as follows:

'For loan upto 10 lakhs the debt equity ratio is 3 : 1. Above 10 lakhs the debt equity ratio is 2 : 1.'[37]

'On need basis the corporation is also providing working capital to the extent of Rs. 10,00,000 under single window scheme. For all types of loan the rate of interest above Rs. 2,00,000 is 16 per cent and below that, the rate of interest is 12 per cent to 13 per cent which is divided by the corporation after scrutinizing the various document along with the project report of the said unit. 25 per cent security margin is to be maintained which may include mortgages of properties, some amount of liquid asset etc. However, in case of financing those items in which the rate of obsolescence or depreciation is very high security margin of 50 per cent is insisted.'[38]

The Corporation at present does not have any separate scheme for financing the fruit and vegetable processing units. However, in the past years the corporation had financed many fruit and vegetable units but the rate of recovery from such units is very poor. Today the Corporation have only one master scheme of financing all the units which also includes the fruit and vegetable processing units. However the management of the corporation strictly see to the viability in the existence and development of the said unit which is to be financed.

Agricultural and Processed Food Products Export Development Authority (APDEA)

The Agricultural and Processed Food Products Export Development Authority (APEDA) was established by an act of parliament passed in December 1985 with its head quarter at New Delhi. The objective of the organisation is to promote and develop

---

[37] Office of the Assam Financial Corporation, Paltanbazar, Guwahati.

[38] *Ibid.*

the exports of various processed products including the fruits and vegetables (fresh and processed) and their products.

APEDA plays a very important role in promotions of agricultural commodities export from the country. It acts as a facilitator and catalyst between exporters and various government agencies and the importers. It helps in product and market development. It provides all information to both exporters and importers, suggest suitable partners for joint ventures, tries to identify new products where India would have competitive advantages.

'In its inception year *i.e.* 1986, 450 exporters from all corners of the country got themselves registered with APEDA.'[39] The number of exporters registered with APEDA increased to 10,000+ in 1996.'[40] Over the years, APEDA has developed a wide database on its scheduled products. The information provided to the registered numbers helps them to decide the line of action in the respective export plans. Moreover, APEDA also acts as a facilitator and takes preactive steps to solve various day to day problems of exports. APEDA also maintains inter government liaison with various ministries *viz.*, civil aviation, agriculture, commerce and also liaison with various nodal agencies *e.g.* State Agro Industries Corporation, Export inspection agencies, customers, Central Board of Excise and Customs etc., for resolving issues pertaining to these areas. APEDA helps the exporters by identifying new markets and products. During the year 1990–91 to 1994–95 Indian agricultural products captured many destinations in the export market. These destinations include UAE, Malaysia, Bangladesh, Indonesia, Russia, Germany, Sri Lanka, Netherlands, Philippines and Oman as well. The exports include many agro based products. By 1994–95 a further range of products got added which included dehydrated, and processed vegetables, mango pulp, processed fruits and vegetables, cereal preparations, walnuts, sheep and goat meat.

Since products like fruit and vegetable, meat, mushrooms and floriculture are highly perishable, proper infrastructure is needed to ensure safety of quality of the product.

---

[39] Office of the Agricultural and Processed Food Products Export Development Authority (APEDA), Assam Industrial Development Corporation Ltd. (AIDC) Campus. R.G. Baruah Road, Guwahati.

[40] *Ibid.*

APEDA has established cold storage facilities at Delhi, Bombay, Bangalore, Calcutta, Madras and Thiruvananthapuram Airports. Efforts are also been made to establish cold storage facilities at the Gopinath, Bordoloi Airport, Borjar, Guwahati.

APEDA has also developed standards of marketing for processed products. The importing countries have certain packaging standards which are to be followed by the exporting country while exporting their products. For this APEDA along with the Indian Institute of Packaging has developed international quality of marketing and exporters using such packaging are provided the benefit of using 'Produce of India' label which serves as a guarantee of packaging to the international importers. A branch office of APEDA is situated at Guwahati which cater the needs of the fruit and vegetable processing units of the entire North east region. Through APEDA many local products of the fruit and vegetable processed units of the North east region which are its registered numbers are exported to various countries. Moreover, APEDA has also organised tours in which local entrepreneurs are taken to the other advanced states outside the north east to get information about the latest techniques of processing fruits and vegetables.

Through APEDA the products are also taken to national and international fairs and exhibitions. It also arranges many seminars and conferences where experts are invited to address the local entrepreneurs regarding the processing of fruits and vegetables and their exports.

Small Industries Service Institute (SISI)

The Small Industries Service Institute (SISI) Guwahati was set up in the year 1951 by the government of India to supplement the activities of the state government for promotion/development of the small scale industries. The Network of the Small Industries Service Institute in the states of Assam, Meghalaya and Arunachal Pradesh is the part of the national network under the Small Industries Development Organisation (SIDO) which is an apex body administered by the Ministry of Industry.

The institute provides assistance and renders various extension services for the growth and development of small scale ventures in the states of Assam, Meghalaya and Arunachal Pradesh in the North eastern region of India. The wide range of services offered by the institute are as follows:

Technical Consultancy Services

The institute provides technical consultancy services to existing industries and new entrepreneurs in the fields of mechanical engineering, chemical engineering, electrical engineering, glass and ceramics, leather foot wears and allied products, food technology and processing, market information etc. The institute also assists entrepreneurs in product selection, choices of technology, selection of machinery etc.

Preparation of Project Report

The institute prepares feasible and viable project report on various items on subsidised charges fixed by the government of India.

Management Development Services

The Small Industries Service Institute renders management consultancy services in important areas like general management, personnel management, financial management, book keeping and accounts, marketing management, cost reduction, quality controls, inventory management, work study and productivity.

Economic Information Services

The economic investigation division conducts studies on economic aspects of small industries for the benefit of existing and prospective entrepreneurs. The studies relate to industrial potentiality of the districts, industrial prospects of products, market surveys, feasibility studies, products reserved for development in small scale sector, feasibility study of industrial estates, monitoring about the SSI units, periodical census surveys of SSI units,

Marketing Services

The Small Industries Service Institute under central government, stores purchase programmes and assists the National Small Industries Corporation Limited to enlist small scale industries for supply of different commodities to government.

Sub Contract Exchange

The institute provides information, enlists and assists small units in obtaining contracts for medium and large scale units for parts/components/sub assemblies etc.

Ancillary Development

Modernisation of selected small scale industries is also covered within the objectives of Small Industries Service Institute. The objective of modernisation programme is to strengthen the small scale units by improving quality and reducing cost. In plant studies are conducted by SISI with the help of specialised professional consultants to draw up modernisation programmes for selected small scale units.

Revival of Sick Units

The Small Industries Service Institute conducts diagnostic in-plant studies of sick units for suggesting ways and means for their revival through the state level co-ordination committee for sick/closed SSI units.

Training Service

Training facilities are provided to upgrade the skills and quality for personnel engaged in the small scale industries. The institutes organises and conducts the specialised courses on financial management and accounting, production management, marketing management, export management, packaging, quality control, inventory management etc. Various courses are also conducted with a view to identify, motivate and train prospective entrepreneurs to set up industries. Moreover, the institute also provides training to industrial workers and artisans to upgrade their skills.

Like in every sector, the SISI also provides its helping hand in the development of the fruit and vegetable processing units of the state. In this context, the SISI provides technical consultancy services and also prepares project reports for the fruit and vegetable processing units.

The small industries mostly provides training facilities to the entrepreneurs of the fruit and vegetable processing units. The SISI conducts EDP programmes in all fields including fruit and vegetable processing. In such programmes, they impart training for producing processed fruit and vegetable products in a most hygienic way. The training is imparted by experts called from the Food and Nutrition Department, Government of India.

They also work as a economic analyst. Analyses are made regarding the potentiality of a particular area for setting up manufacturing units making a particular product and provide such information to the entrepreneurs when required.

District Industries Centre (DIC)

The District Industries Centre was started in the year 1979–80 under Department of Industries, Government of Assam. Since then the centre has been working for setting up and developing of the local industries. Initially there were altogether five district industries centre in the whole of Assam. These were at Guwahati, Dibrugarh, Silchar, Nagaon and Sibsagar. At present all the 23 districts of Assam has a District Industries Centre. The respective District Industries Centres provide various financial assistance to the units located in the respective districts.

Each district industries centre has a General Manager as its head, followed by financial manager and project manager, assistant manager, extension officer and the office staff.

The District Industries Centre is not a financial institution. It does not provide financial assistance directly to the units but recommend the units to the bank for providing finance after studying their viability.

The centre provides assistance to the SSI units, Tiny units small scale service and business enterprises (SSSBE's), sick units/relief undertakings subject to a maximum period of 3 years, unit set up by women entrepreneurs, large and medium units, export oriented units, units undergoing expansion, diversification, modernisation.

The various incentives provided are as follows:

Power Subsidy

'50 per cent power subsidy is granted upto 1 Mega watt, the maximum ceiling of which is Rs. 5 lakhs per industrial unit per year. The amount of subsidy is 30 per cent above 1 mega watt and upto 5 mega watt the ceiling of subsidy being Rs. 15 lakhs per industrial unit per year. Again the amount of subsidy is 20 per cent above 5 mega watt the ceiling of subsidy being Rs. 30 lakhs per industrial unit per year.'[41]

---

[41] Office of the District Industries Centre, Kamrup, Assam.

The power subsidy is available for a period of five years from the date of commercial production. This subsidy is made available in the form of reimbursement of fully paid power bills.

Interest Subsidy on Working Capital

'5 per cent interest subsidy is provided to the SSI units with an investment upto Rs. 60 lakhs on interest in working capital loan obtained from the banks and financial institutions. This benefit is provided for a period of 3 years from the date of commercial production and the maximum benefit is Rs. 3 lakhs per year per unit.'[42]

State Capital Investment Subsidy

'A special state capital investment subsidy at the rate of 30 per cent of the capital investment on land, building and plant and machinery etc subject to a ceiling of Rs. 10 lakhs is provided to the industries under this policy.'[43]

Sales Tax Exemption

Sales tax exemption is granted to all new and existing units going in for expansion, diversification and modernisation for sale of finished products and purchase of raw materials. For new units set up under the small scale, Tiny and the small scale service and business enterprise (SSSBEs) sector sales tax exemption is granted upto 7 years subject to maximum of 150 per cent of fixed capital investment. For new units in the medium and large scale sector it is granted upto 7 years subject to maximum of 100 per cent of fixed capital investment. For expansion, diversification and modernisation of the SSI, tiny and SSSBEs units sales tax exemption upto 7 years subject to maximum of 100 per cent of additional fixed capital investment. For expansion, diversification and modernisation of the medium and large scale units exemption is given upto 7 years to maximum of 90 per cent of additional fixed capital investment'.[44]

---

[42] Office of the District Industries Centre, Kamrup, Assam.

[43] *Ibid.*

[44] *Ibid.*

'For the SSI, Tiny and SSSBE'S units which are sick sales tax exemption granted upto 3 years subject to maximum of 100 per cent of additional investment made for rehabilitation. In case of large and medium scale units such exemption is granted upto 3 years subject to maximum of 100 per cent of additional investment made for rehabilitation. In case of Electronic industries the tax benefit is upto 250 per cent of fixed capital investment spread over a maximum period of 7 years in view of low fixed capital investment. The period of exemption is limited to 7 years which is proposed to be extended upto 10 years synchronising with the North East Policy.'[45]

Subsidy on Generating Set

'The subsidy on the generating set including non conventional energy generating sets is given at the rate of 50 per cent of the cost of the generator subject to a ceiling of Rs. 10 lakhs per industrial unit.'[46]

Contribution to Feasibility Study Cost

For large and medium scale units the cost of feasibility report prepared by agencies approved by Udyog Sahayak of AIDC/Directorate of industries will be subsidised to the extent of 90 per cent subject to a ceiling amount of Rs. 2 lakhs in each case. The contribution will be treated as interest free loan for a period of five years from the date of commercial production or from the date of disbursement of the loan whichever is later. If the project is not implemented within the prescribed period, the feasibility report shall be the property of AIDC/Directorate of industries and the entrepreneur shall be liable to pay back the entire amount to AIDC within a prescribed time limit.' For small scale units the cost of the feasibility report prepared by an agency approved by the Udyog Sahayak will be subsidised to the extent of 100 per cent in case of projects whose total project cost is within Rs. 10 1akhs and 90 per cent in case of projects above 10 1akhs, the ceiling on subsidy being Rs. 50,000 in each case.'[47] The feasibility report will become the property of the government if the project is not implemented within the prescribed time.

---

[45]   Office of the District Industries Centre, Kamrup, Assam.

[46]   *Ibid.*

[47]   *Ibid.*

Subsidy on Infrastructure Facilities

In appropriate cases the centre allots land on hire purchase basis to the entrepreneurs. The cost of the land including cost of development and cost of creation of infrastructural facilities like power, water and approach road will be recovered in annual instalments over 15 years from the date of handing over of the land with a moratorium of five years, such land will be permitted to be utilised by the unit for mortgage and hypothecation for obtaining loans from banks and financial institutions. In case of developed land not available for allotment the entrepreneurs are allotted undeveloped lands. In such situation actual land development cost is provided on interest free loan to the eligible units subject to certain ceilings. 'If the cost of the project is upto Rs. 2 crores the limit of interest free loan is 3 per cent of the project cost the maximum limit of such interest free loan is Rs. 5 lakhs.'[48]

'For project cost above Rs. 2 crores and upto Rs. 5 crores the interest free loan is 2 per cent of the project cost, the maximum limit of such loan is Rs. 7.50 lakhs. For project cost above Rs. 5 crores and upto Rs. 10 crores the interest free loan is 2 per cent of the project cost, the maximum ceiling of such loan is Rs. 15 lakhs. In case of projects the cost of which is above 10 crores and upto Rs. 50 crores the interest free loan is 1.5 per cent of the project cost, the maximum limit of such interest free loan is 50 lakhs. In case of project above Rs. 50 crores the interest free loan is 1 per cent of the project cost, the maximum limit of such loan is Rs. 100 lakhs.'[49]

Manpower Subsidy

'Moreover subsidy on manpower development is also provided in respect of local persons trained and employed in the industrial units. For managerial category the amount on training is Rs. 10,000 per person, for supervisory category the amount spent on training is Rs. 7,500 per person for skilled category the amount spent on training is Rs. 5,000 per person and in case of unskilled person the amount spent on trainings Rs. 2,000 per person.'[50]

---

[48]   Office of the District Industries Centre, Kamrup, Assam.

[49]   *Ibid.*

[50]   *Ibid.*

'For unit with investment upto Rs. 2 crores the total ceiling of the amount spent on training is Rs. 1 lakh. For units with investment above Rs. 2 crores and upto Rs. 5 crores the total ceiling is Rs. 2 lakhs, above Rs. 5 crores and upto Rs. 10 crores it is Rs. 5 lakhs and above Rs. 10 crores the total ceiling is Rs. 7 lakhs.'[51]

Special Incentives to the Pioneer Unit

'In case of pioneer unit the centre gives an additional state capital investment subsidy of 5 per cent of fixed capital investment subject to a ceiling of Rs. 10 lakhs. Such units will also be given power subsidy for an additional period of 2 years at the same rate in addition to power subsidy at 5 : 1.'[52]

Special Incentives for Export Oriented Units (EOU'S)

'The centre provides special incentives for 100 per cent export oriented units. It provides additional state capital investment subsidy of 10 per cent subject to a ceiling of Rs. 10 lakhs. Moreover, it also provides additional 20 per cent subsidy on purchase of testing equipment for obtaining ISO 9000/BIS 14000 series registration subject to a ceiling of Rs. 2 lakhs. Further the centre provides a subsidy on the purchase of testing equipment for obtaining ISO 9000/BIS 14000 series registration at the rate of 30 per cent of the cost of the equipment subject to a ceiling of Rs. 5 lakhs.'[53]

Special Incentives for Women Entrepreneurs

'Special incentive are granted to women Entrepreneurs by the centre. This includes additional state capital investment subsidy of Rs. 5 lakhs where women constitute more then 50 per cent of the workforce in the industry. Moreover, additional subsidy on working capital is given having interest 2 per cent for a period of three years from the date of going into commercial production subject to a ceiling of Rs. 1 lakh per annum. Further, factory sheds are allotted to women entrepreneurs on priority basis on a subsidised rent at the rate of 75 per cent of the economic rent for a period of five years from the date of going into commercial production.'[54]

---

[51]   Office of the District Industries Centre, Kamrup, Assam.

[52]   *Ibid.*

[53]   *Ibid.*

[54]   *Ibid.*

Special Incentive for Physically Handicapped Person

'Special incentive are provided to physically handicapped person which includes additional 5 per cent state capital investment subsidy subject to a ceiling of Rs. 5 lakhs. Moreover, additional 2 per cent investment subsidy on working capital for a period of three years from the date of commercial production subject to a ceiling of Rs. 1 lakh per annum.'[55]

Special Incentives for Handicraft Industries

'The centre provides special incentives for handicraft industries. It provides additional 10 per cent special capital investment subsidy subject to a ceiling of Rs. 2 lakhs, and additional 2 per cent interest subsidy on working capital for a period of three years from the date of going into commercial production subject to a ceiling of Rs. 1 lakh per annum.'[56]

Miscellaneous Subsidy

'The centre also provides certain miscellaneous subsidy to the eligible units. This includes 20 per cent of the cost payable to Assam State Electricity Board for drawal of High transformer/Low transformer line upto the premises of the unit and installation, of transformer for power supply to the unit subject to a ceiling of Rs. 1 lakh, 50 per cent of the fees (excluding recurring loyalty) paid for procurement of know how from national research and development corporation or other agency recognised by Udyog Sahayak subject to a ceiling of Rs. 1 lakh. Moreover, 50 per cent of the cost of pollution control and monitoring equipment subject to a ceiling of Rs.2 lakhs. Further, the centre also provides 50 per cent of the cost of quality equipment subject to a ceiling of Rs. 1 lakh. The miscellaneous subsidy shall be released only after actual expenditure is made by the unit. These incentives are applicable to all sectors of industries where the fixed capital investment does not exceed Rs. 5 crores.'[57]

Prime Minister's Rozgar Yojana

'Prime Minister's Rozgar Yojana popularly known as PMRY is another loan scheme of the government of India for providing self

---

[55] Office of the District Industries Centre, Kamrup, Assam.

[56] *Ibid.*

[57] *Ibid.*

employment to unemployed youth which is operated through the district industries centre. The loan is available for ventures whose total project cost is within Rs. 2 lakhs for industry and other activities and Rs. 1 lakh for business in case or individuals. The entrepreneurs are required to contribute 5 per cent to 12.5 per cent of the project cost. The repayment schedule ranges from 3 to 7 years after initial moratorium of 6 to 18 months.'[58]

The district industries centre is the implementing agency for PMRY. The centre in consultation with the banks of the respective areas are responsible for the formulation of self employment plans, their implementation and monitoring under the over all guidance of the district PMRY committee. A district level task force committee is also constituted which comprise officials from SISI, Banks, employment exchange which are responsible for motivation and selection of beneficiaries, identification and preparation of the schemes. The task force recommends loan applications to the banks for getting speedy clearance.

Assistance given to the Fruit and Vegetable Processing Industry

'There is no separate scheme for providing assistance to the fruit and vegetable processing industries. The general schemes which apply to all the other industries are also applicable to the fruit and vegetable processing industries including the PMRY scheme. However, in case of agro and food processing industries an additional state capital investment subsidy of 5 per cent subject to a ceiling of Rs. 5 lakhs is provided. Moreover, 50 per cent of the cost payable for getting FPO licences/Agmark/Trade mark for the products for food processing industries subject to a ceiling of Rs. 1 lakh is also provided.'[59]

National Small Industries Corporation Limited (NSIC)

The National Small Industries Corporation was established in the year 1955 by the government of India with a view to promote and foster the growth of small industries in the country. NSIC remains the forefront of industrial development throughout the country with

---

[58] Office of the District Industries Centre, Kamrup, Assam.

[59] *Ibid.*

its various programmes and projects to assist the small scale sector in the country. The small scale sector continues to remain an important instrument for enterprise building, dispersal of industries for even regional development and employment generation. NSIC has been successfully able to play its assigned role in this endeavour. Due to the changed industrial scenario and gradual globalisation of the economy small scale sector has to face a stiff competition, NSIC enable the small scale sector to meet this challenge by initiating various steps. The National Small Industries Corporation provides diversified support by providing several assistance to the small scale sector. Basically the corporation provides equipment and marketing to the SSI units by hire and purchase and equipment leasing scheme. The quantum of finance is minimum 1 lakh rupees and maximum 1 crore rupees. However, in case of women entrepreneurs the minimum amount of finance is fifty thousand rupees. The rate of interest is 14 per cent and the payment should be made quarterly in 20 equal instalments within a period of five years. The gestation period is 6 months. The supplier from whom the machinery is procured can be selected by the entrepreneur himself provided the former is registered with the NSIC. Besides this NSIC also provides raw material assistance upto ten thousand rupees at a time, Moreover, NSIC also helps the small scale units to sell their goods and services to the government departments and agencies through single point registration scheme. The NSIC enlists small scale units as competent to undertake supply of various items to the government. The single point registration scheme is to avoid multiplicity of registration with various government agencies and to ensure that the units registered with NSIC are considered at par with those registered directly with the purchasing agency. Bonafide small scale units registered with the Directorate of industries/District Industries Centres are enlisted under this scheme. Furthermore, the NSIC has provided various marketing support programmes by setting up marketing development centres which provides marketing outlets and serve as a common marketing facility for a number of small scale units. NSIC has recently started an integrated marketing support programme in which bills pertaining to supplies made by small scale units to eligible purchases are discounted by NSIC up to certain specified limit. The scheme has been introduced with a view to mitigate the problem of delayed payment by buyers against supplies made by small scale units. NSIC has helped the SSI units in technology upgradation by providing proper guidance and latest information in connection with technology upgradation and dissemination among the other small and large scale enterprises spread all over India and abroad. Further NSIC is also engaged in exporting a number of items manufactured by SSI units. The corporation provides a complete package of export assistance, testing facilities, pre shipment, credit facility, export incentives etc., apart from exposure to the products of small scale units in trade fairs, buyers and sellers meet etc. In addition to this, NSIC also helps the units in procurement of samples, development of counter samples and in negotiations for first business deals with foreign buyers.

Assistance to the Fruit and Vegetable Units of Assam

Though NSIC provides many assistance for procuring of machinery and equipment, but regarding fruit and vegetable processing units the corporation has not yet provided assistance to any such unit in the state because as the fruit and vegetable units are mostly unorganised the NSIC hesitates to grant loan to such units out of the fear that the loan shall not be refunded. Moreover, since the units are mostly unorganised with very little own contribution they could not fulfil the strict criteria of the NSIC for granting loan.

Further, in case of assistance with regard to supply of raw materials the NSIC finds quite difficult in providing raw material assistance to the units because the payment of such supply is made on the name of the supplier who supplies the raw materials and who should be a registered one with the NSIC. This is not possible in case of fruit and vegetable units because there is no organised supply or supplier of raw materials in this sector. Moreover the prices of raw materials fluctuate frequently.

Indian Institute of Entrepreneurship (IIE)

Indian Institute of Entrepreneurship (IIE) was established in the year 1993 by the Ministry of Industry, Government of India with its head quarter at Guwahati. The institute was registered on March 1993 under the societies registration act 1860. In April 1994 it took over the North East Regional Centre of NISIET, Hyderabad and started its activities from its premises. The management of the institute is in the hands of a Board of management appointed by the government of India. The president of the institute is the chairman of the North Eastern Council. The institute undertakes training, research and consultancy activities in the field of small scale industry and the entrepreneurship. The activities of the institute include identification of training needs, designing and organising training programmes both for development functionaries and entrepreneurship, evolving effective training strategy and methodology for different target groups and locations, organising seminars, workshop and conferences for providing forum for interaction and exchange of views by various agencies and entrepreneurs, undertaking research on entrepreneurship development, documenting and dissemination of information needed for policy formulation and implementation on self employment and entrepreneurship.

The institute also acts as a catalyst for entrepreneurship development by creating an environment for entrepreneurship in the support system, developing new entrepreneurship, helping in the growth of existing entrepreneurs, encouraging the educated sons and daughters of artisans to take up their family business and also by taking up programmes on entrepreneurial education. The institute takes up these activities for creating an entrepreneurial culture in the society. Most of the programmes organised by the institute are sponsored by the Government of India, North Eastern Council, SIDBI (Small Industries Development Bank of India) and also by international agencies.

The institute has been organising general, target group oriented, product specific and location specific entrepreneurship development programmes (EDP's) for creating new entrepreneurs. There are EDP's for women entrepreneurs, rural entrepreneurs, product specific EDP's for food processing industries, jute products and other agro products.

The institute have also taken effort to create interest among the educated sons and daughters of the artisans particularly in the handicraft sector to save these units from getting out of existence.

Moreover in order to orient and sensitize the environment for entrepreneurship, the institute has been organising training programmes, workshops and seminars. It organises entrepreneurs and appraising projects in the SSI sector in collaboration with EDI, Ahmedabad and combined programmes both for bankers, DIC functionaries and also development functionaries. Besides this, it also organises a member of district, state and regional level workshops on sharing experience on implementation of PMRY.

The institute also imparts a three month's certificate course on Entrepreneurship and a Entrepreneurial awareness camp for school students.

The institute also organises seminars and workshops for sharing experiences on implementation of programmes of self employment and entrepreneurship. Such workshops and seminars creates awareness about the prospects of developing entrepreneurship in certain line of activity such as handmade paper industry, food processing industry etc.

Therefore the training, research and the consultancy activities carried on by the Indian Institute of Entrepreneurship in the field of small industry and entrepreneurship helps in the growth and development of many local industrial units, including the fruit and vegetable processing units.

The IIE conducts many training programmes, conferences and seminars which help the entrepreneurs of the local fruit and vegetable processing units for developing their technology, marketing, quality improvement etc. These training programmes have also helped in creating new entrepreneurs for fruit and vegetable processing units in the state of Assam.

Small Industries Development Bank of India (SIDBI)

Small Industries Development Bank of India (SIDBI) was set up by an act of parliament as an apex institution for promotion, financing and development of industries in the small scale sector and for co-ordinating the functions of other institutions engaged in similar activities. SIDBI is a wholly owned subsidiary of Industrial Development Bank of India (IDBI). It commenced its operations on April 2, 1990 by taking over the outstanding portfolio and activities of IDBI pertaining to the small scale sector SIDBI is operating through its head office at Lucknow and a network of 5 Regional offices and 28 branch offices in all the states.

SIDBI under the charter has been assigned the task of being the main surveyor of term finance to the small scale sector in the country. Small scale industrial units, artisans village and cottage industrial units in the tiny sector and small road transport operators are extended financial assistance mainly by way of refinance through primary lending institutions (PLIs) *i.e.* the State Financial Corporation (SFCs), State Industrial Development Corporations/State Industrial Investment Corporations (SIDCs/SIICs) and banks which have a wide network of branches. Term loans extended by the primary lending institutions to small scale industrial projects, irrespective of location and form of organisation of the units are eligible for refinance assistance.

With a view to encouraging bills culture and helping the SSI units realise their sale proceeds of capital goods/equipment and components/sub assemblies/intermediates, SIDBI directly discounts bills arising out of these transactions and also rediscounts those bills that are discounted by the banks. In addition, SIDBI has also introduced direct finance schemes for specialised marketing agencies, sub contracting/ancillary units and infrastructure development agencies so as to fill the gaps in these areas in the existing credit delivery mechanism SIDBI has been providing assistance to well run companies on a selective basis for acquiring machinery and equipment both indigenous and imported for modernisation, expansion, diversification, balancing scheme etc under equipment finance scheme (EFS). Project loans are also considered on a selective basis. SIDBI operates a venture capital fund for support to ventures in the small scale sector which have special features in terms of technology, market prospects, return on investment or entrepreneurial profile.

All projects in the small scale sector are normally eligible for assistance. There is no separate scheme for providing assistance to the fruit and vegetable processing units. All schemes applicable to other small scale units are also applicable to the fruit and vegetable processing units provided the terms of the schemes suits the project of the unit. Various schemes of SIDBI can be broadly classified into four categories. They are direct assistance scheme, indirect assistance schemes, refinance scheme and bill discounting schemes. Under the direct assistance scheme, finance is provided for setting up of new units and expansion, diversification, modernisation of existing units. The project cost should not be less than 75 lakhs. The various schemes under the direct assistance are project finance scheme, Venture capital scheme, scheme for financing activities relating to marketing of SSI projects, scheme for ancillary units, schemes for development of industrial areas for SSI sector, Equipment finance scheme, scheme for integrated infrastructural development, ISO 9000 scheme, scheme of assistance for savings cum credit Groups (Micro credit scheme), scheme for foreign currency term loans to SSI units, PCFC loans to small scale industries, scheme for opening of foreign letters of credit, Technology development and modernisation fund scheme.

Indirect assistance is provided through the banks, SFCs and SIDCs. There is no minimum loan limit for such finance. In case of refinance there are several schemes like the general scheme for setting up of new units and expansion, modernisation and diversification of existing units, schemes for cottage and village industries, equipment refinance scheme, schemes for small road transport operators, schemes for qualified professionals in management, accountancy, medicine, architecture, engineering etc, schemes for tourism related activities, schemes for hotel and restaurant projects, schemes for infrastructure development, equity type assistance scheme to meet the gap in minimum promoters contribution and in equity, schemes for women entrepreneurs, special schemes for assistance to ex-servicemen, single window scheme, refinance scheme for technology development and modernisation, refinance scheme for acquisition of ISO 9000 series certification by SSI units.

The last category of schemes are the bills schemes which include the bills rediscounting scheme and the direct discounting scheme. The minimum amount of project cost varies according to schemes. 'The minimum promoter's contribution prescribed under the schemes generally varies between 10 per cent and 25 per cent. The debt equity ratio upto 3 : 1 is normally applicable under all refinance schemes in respect of loan amount upto Rs. 10 lakhs and upto 2 : 1 above Rs. 10 lakhs. Interest rate for most activities is related to the size of loan under various schemes of assistance. Repayment period for the term loan is fixed after taking into account the profitability and debt servicing capacity of the project.'[60]

National Horticulture Board (NHB)

National Horticulture Board (NHB) was set up by the Government of India in 1984 as an autonomous society under the

---

[60] Regional Office of the Small Industries Development Bank of India (SIDBI). G.S. Road, Guwahati.

societies registration Act, 1860 with a mandate to promote integrated development of horticulture, to help in coordinating, stimulating and sustaining the production and processing of fruits and vegetables and to establish a sound infrastructure in the field of producing, processing and marketing with a focus on post harvest losses. The head office of NHB is at Gurgaon, Haryana. At present there are 33 offices of NHB functioning throughout the country. The organisational structure of NHB consist of Board of Directors, Managing Committee and principal executive.

The broad objectives of NHB are as follows:

1. Develop high quality horticultural farms in identified belts and make such areas vibrant with horticulture activity which in turn will act as rules for developing commercial horticulture.

2. Develop post harvest management infrastructure.

3. Strengthen market information system and horticulture database.

4. Assist research and development programmes to develop products suited for specific varieties with improved methods and horticulture technology.

5. Provide training and education to farmers and processing industry personnel for improving agronomic practices and new technologies.

6. Promote consumption of fruits and vegetables in fresh and processed form.

The various schemes of NHB are as follows:

Development of Commercial Horticulture through Production and Post Harvest Management

This scheme is of two components:

(a) Production related

(b) Post harvest management/processing related.

'The pattern of assistance is in the form of back ended capital investment subsidy not exceeding 20 per cent of the total project cost with a maximum limit of Rs. 25 lakh per project. For north eastern Tribal/Hilly areas maximum limit of subsidy is Rs. 30 lakhs per project. Subsidy would be sanctioned and released through participating banks and financial institutions. 50 per cent of the subsidy would be released to the leading participating financial institutions/NABARD/Bank/financial institution after the loan is sanctioned and 50 per cent of the loan is disbursed and remaining 50 per cent on completion of the projects.'[61] The financial institution and banks shall include NABARD, IDBI, SIDBI, ICICI, state financial corporation, State Industrial Development Corporation other designated loaning institution of the states/Union/ Commercial/ Coopertive banks etc.

Capital Investment Subsidy for Construction/Expansion/ Modernisation of Cold Storage/Storages for Horticulture Produce

The aim of this scheme is to promote the setting up of cold storages in the country for reducing post harvest losses. The cold storage include controlled atmosphere (CA) and modified atmosphere (MA) stores, pre-cooling units and other storage for onion etc.

'The pattern of assistance includes 25 per cent promoter's contribution, 50 per cent term loan by bank and 25 per cent back ended capital invetsment subsidy by NHB.'[62]

Technology Development and Transfer for Production of Horticulture

The main aim of this scheme is to popularize new technologies, tools and technologies for commercialisation, introduction of new concepts to improve farming system, upgradation of skills by exchange of technical know how, identification and popularisation of indigenous crops with emphasis on domestic and export promotion.

The eligibility and pattern of assistance is different for different components.

For introduction of new technologies, grant upto 100 per cent of the cost is provided under this scheme. However, this is restricted upto Rs. 10 lakhs in case of pilot project based on high quality

---

[61] Office of the National Horticulture Board (NHB), Guwahati 28.

[62] *Ibid.*

commercial production, compact area development approach for popularizing new and modern scientific concepts in horticulture and Rs. 25 lakhs for project falling under post harvest management, primary processing, Biotechnology and introduction of new equipment and machinery and research and development project. Moreover in case of national visit of progressive farmers financial assistance in the form of grant in aid shall be provided for organising such visit or training upto a period of 10 days (Excluding the journey). 'The assistance also includes second class rail fare plus boarding charges @ Rs. 100 for state level participation, Rs. 5 lakhs for national level participation and upto Rs. 10 lakhs per event for international participation.'[63]

'The duration of the seminar for the above mentioned assistance should be 3 to 5 days. In case of short duration that is 1 to 2 days financial assistance upto Rs. 1 lakhs for state level participation, Rs. 2 lakh for international event NHB also provides assistance for publication and preparation of films relating to horticulture. Besides this, NHB also organises study tours for its officers to give exposure to the officers about horticulture industry in advanced countries. Further a grant upto Rs. 20,000 for one expert for a group of upto 5 experts per project is provided with a ceiling of Rs. 1 lakhs per project for honorarium to scientist for facilitating effective transfer of technology as per the details of schemes covered under development of high quality commercial horticulture and transfer of technology.'[64]

Establishment of Nutritional Gardens in Rural Areas

The main object of this scheme is to improve nutritional status of the rural masses and to increase the per capita availability of fruits and vegetables. Moreover, it also insists in making the availability of fresh fruits and vegetables throughout the year and preservation of horticultural produce. Under this scheme NHB provides 100 per cent grant in aid.

Market Information Service for Horticulture Crops

NHB also provides latest information about the daily price of horticulture crops throughout the country. At present there are 33

---

[63] Office of the National Horticulture Board (NHB), Guwahati 28.

[64] *Ibid.*

Market information centre which collects the information on prices and arrivals in their respective markets and communicate it to the Central Coordinating cell of NHB Head quarter at Gurgaon which complies and analyses the prevailing prices and arrivals of selected horticulture commodities of commercial importance.

Horticulture Promotion Service

This scheme relates to the review of the present situation of horticulture development in particular area/state. It includes the development of primary and secondary data of various aspects on horticulture. The scheme further includes identification of constraints and suggest their remedial measures, development of short term and long term strategies for systematic development of horticulture and providing for consultancy services in pursuance there of. The pattern of assistance includes 100 per cent cost of study to be borne by the board.

The eligible promoters under the above schemes shall include NGO'S, Association of growers, Individuals, Partnership, Proprietory firms, Companies, Corporations, Cooperatives, Agricultural produce Marketing Committees, Marketing Boards/ Committees, Municipal Corporations/Committees, Agro industries Corporation and other connected Research and Development organisation. However, individuals and other research and development organisations are not eligible for the cold storage capital subsidy scheme.

North Eastern Industrial and Technical Consultancy Organisation Limited (NEITCO)

NEITCO is a consultancy organisation sponsored by Industrial Development Bank of India (IDBI), Industrial Finance Corporation of India (IFCI) Industrial Credit and Investment Corporation of India (ICICI), Nationalised Bank and State level development and financial corporations. NEITCO is one amongst a family of seventeen (17) Technical Consultancy Organisations (TCO's) set up in the country by the financial institutions to provide complete consultancy and escort services to tiny, cottage, small scale and medium scale industries.

NEITCO is the second TCO to be set up in the country in 1973 with its headquarter at Guwahati and branch offices in other states.

NEITCO is a Government limited company registered under this Companies Act 1956, and headed by a chairman (Non executive) with the day to day affairs being administered by a Managing Director. NEITCO comprises of a band of professionals from various disciplines like MBA's, Economists, Trainers, Finance and Computer professionals. With its band of professionals NEITCO carries various activities like project formulation, motivational training project implementation and market survey. Apart from the head office at Guwahati there are two branch offices one each at Shillong and Itanagar which were established in August 1988 and June 1989 respectively.

Main Objectives

The main objectives of NEITCO are as follows:

1. To identify industrial potential through surveys and otherwise.

2. To prepare project profiles, feasibility reports and pre-investment studies in respect of specific industries.

3. To identify potential entrepreneurs for implementation project and provide technical and administrative assistance to them for promotion and management of industries.

4. To undertake techno-economic appraisal of projects.

5. To undertake research and surveys for specific products.

6. To act generally as an industrial management and financial consultant.

7. To undertake project supervision and where necessary render technical and administrative assistance for improving the working of industrial concerns.

8. To undertake any type of research and service in order to promote the objectives of the company for evaluating or dealing with marketing or investments and to undertake and carry on techno-economic or other studies or surveys in connection with the development of industries.

NEITCO has four main divisions:

1. Consultancy division
2. EDP division

3. Project division

4. Rural development division

Consultancy Division

The consultancy division is responsible for providing technical and management consultancy services. It provides various assignments such as project identification/project profiles, Market surveys, diagnostic studies, Industrial potential surveys and planning of industrial areas, growth centres. The above assignments cover a range of industrial sectors such as cement, Textiles, food processing Electrical/Electronics and chemicals.

EDP Division

Entrepreneur Development Programme (EDP) activities commenced on a regular basis in 1984. Four types of EDP'S are being conducted which are General EDP'S, science and technology (S and T) EDP'S, women EDP'S and rural EDP'S. These EDP'S differ mainly with regard to the target groups. The science and technology EDP'S are meant for science and technology graduates/diploma holders, the women candidates and rural EDP'S are oriented towards candidates in rural areas. Over 2000 candidates have been trained and motivated during the last seven years and of these over 700 entrepreneurs have successfully set up and are operating their units.

Project Division

The project division has been involved in implementing projects from concept to completion. The projects implemented include cement plants, Dye-houses, Tantalum capacitor project and a fruit juice concentration plant.

Rural Development Division

NEITCO has also carried out employment generation programmes targeted at those with minimal education such as school, college drop outs. Such candidates are given training in technical trades, such as plumbing, masonry, automobile, repair, electrical wiring as also in areas like weaving, poultry/piggery farming, fishery, mush room cultivation. After training, necessary assistance are given which includes tie up for finance from institutions, banks to enable them to be self employed or tie up with establishment where wage employment is available. Moreover, block adoption programmes are carried on to identify need biased income generating activities that would catalase economic development of the village in a specific block.

Role of NEITCO in the Fruit and Vegetable Processing Sector

As stated earlier the central activity of NEITCO is to draft project reports, carry out market survey, location survey, availability of raw material survey in a particular area etc upon which the viability of a particular business in that area depends.

With regard to these NEITCO performs the same type of activities in the field of fruit and vegetable processing also. These activities include drafting of project report, survey of horticultural products, market surveys and also EDP's on food processing etc.

Though the Governmental and Semi Governmental institutions have taken up many schemes for the upliftment of the local fruit and vegetable based units yet according to the entrepreneurs practically such assistance are not fully implemented and are not satisfactory. This is proved by the hypothesis no. 2 which states that the existing Governmental support to the fruit and vegetable based units are not up to the mark. The hypothesis is proved through the opinion of the entrepreneurs which is secured by the distribution of questionnaire among them. The information gathered is shown in the Table 6.1 in the form of number and percentage.

Hypothesis No. 2

That the existing governmental support to the fruit and vegetables based units are not up to the mark.

Table 6.1 showing government assistance to the fruit and vegetable units upto 1999.

The Table 6.1 shows the information gathered from the survey work conducted among the entrepreneurs of the fruit and vegetable processing industry about the existing government support received by the units. Out of the 87 units registered with the Directorate of Industries, only 50 per cent of such were covered during the field study. This is because some of the units are closed down and location of some units could not be traced.

Table 6.1: Opinion of Entrepreneurs Relating to Receipt of Government Assistance

Types of Queries	Opinion of Entrepreneurs			
	Yes		No	
	Number	Percentage	Number	Percentage
Shed allotted by the government	4	9%	39	91%
Government help in promoting sales	12	27%	31	73%
Government facilities for preservation of raw materials and finished products	–	–	43	100%
Government assistance for packing of the finished products	–	–	43	100%
Government assistance for exporting of the local products	4	9%	39	91%
Whether received any benefit from the training programmes conducted by the government	19	45%	24	55%

Source: Field study.

The Government support as per the table is not sufficient enough. Out of the units surveyed only 9 per cent have been provided with sheds by the government. But according to the entrepreneurs of those units provided with government sheds the space area of such sheds are not sufficient to carry out the production and other activities. This is shown in Table 6.2.

Table 6.2: Opinion of Entrepreneurs in Percentage Regarding Space Allotted to the Units by Government

Types of Queries	Opinion of Entrepreneurs	
	Sufficient	Insufficient
Space allotted by Government	Nil	100%

Source: Field study.

Moreover, Government assistance for promoting sales have been received by only 27 per cent of the units surveyed of which as per Table 6.3 majority have regarded Government assistance for promoting sales as profitable but most of them cannot avail it due to the strict and hard criteria which has to be fulfilled for getting such assistance.

Table 6.3: Opinion of Entrepreneurs in Percentage Regarding the Sale of Products through Government Assistance

Types of Queries	Opinion of Entrepreneurs	
	Profitable	Not Profitable
Sales through Government assistance	66%	34%

Source: Field study.

Again as per Table 6.1 none of the units surveyed have received government assistance for preservation of raw material and finished products nor any of them have received any assistance for packing of the finished products. Further, as per Table 6.1, only 9 per cent of the units surveyed have received government assistance for exporting their products. Out of them majority are of the opinion that such assistance is not sufficient. This is shown in Table 6.4.

Table 6.4: Opinion Entrepreneurs in Percentage Regarding the Government Assistance Given for Export of Products

Types of Queries	Opinion of Entrepreneurs	
	Sufficient	Insufficient
Government assistance for exporting of products	25%	75%

Source: Field study.

Furthermore, though majority of the units surveyed have received financial assistance from the Government yet all of them are of the view that such assistance is unsatisfactory. This is shown in Table 6.5.

Table 6.5: Opinion of Entrepreneurs in Percentage Regarding Financial Assistance Received from Government

Types of Queries	Opinion of Entrepreneurs	
	Satisfactory	Unsatisfactory
Financial assistance received from Government	Nil	100%

Source: Field study.

So from the above findings the hypothesis that the existing government support to the fruit and vegetable processing units are not up to the mark can be accepted.

Chapter 7

Special Problems of Marketing of Fruit and Vegetable Based Units in Assam

Packaging and Labelling

Packaging

Packaging can be defined as the activities of designing and producing the container or wrapper of a product. Most of the physical products are sold in packages. Packaging plays a very important role in today's marketing world. A beautiful package or wrapper of a product helps in attracting customer towards the product. Therefore packaging has been rightly described as the silent salesman.

According to the Indian Institute of Packaging, 'It is the embracing function of package, selection manufacture, filling and handling.'

Importance of Packaging

In recent times packaging has become an important marketing tool. Modern methods of packaging are valuable to the manufacture to establish his branded products as distinct from those of his rivals. The more effectively a product is packaged the more effective is its identity and individuality. Packaging is advertising on the shelf a means of attractive display in the retailer's shop. Thus a good package ensures ultimate success of the product. It provides an invaluable aid to decision making by the customers. It also acts as an important information clue to the buyers. Packaging helps the consumer and establish preferences that ensures reported repurchase of a particular product.

In the present age of consumer oriented marketing approach packaging has gained unique importance. The utility reasons for packaging that is protection, identification and convenience are themselves exploited in selling and some features of the packages may serve as a sales appeal, for example, reusable jar packaging decorates and beautifies the product so as to lead the implusive buying. Packaging also serves as a device of sales promotion.

Packaging is essential for the safe and easy marketing of any food products and for the retention of their natural characteristics till consumption or utilisation otherwise. Food packaging is an integral part of food processing. Today consumers are very much conscious about the outside package of the product. They are seen to be much more attracted towards the beautifully packaged products. Since food is directly connected with the health therefore the manufactures and the packer must be over conscious while packing the products. Besides giving an attractive outward look, packaging also protect the products especially the food items from getting contaminated. That is why manufacturing and packaging of the food products should be done in a clean and hygienic manner.

Packaging Cost

Packaging cost can vary from one country to another depending on the availability of raw materials, cost of local manufacturer, degree of protection provided against competition, imports etc. Indian consumers are accustomed to buying products loose for which they bring their containers. They believe that expenditure or elaborate packaging is a waste. But over a period of time consumers buying habits have changed and modern consumers look forward to easy to handle, convenient to take home and eye catching packs. On the other hand, manufactures had to pay much on T.V., radio, printing media for sales promotion. As a result, at times the cost of packaging is as much as the cost of the contents. To reduce the high cost of packaging recycling of containers or reuse or glass bottles are much in use.

Recent Development of Packaging

Over the years, great changes have taken place in packaging materials. In the earlier days, wood was the main material used for packing. But gradually due to the shortage in the supply of wood this was replaced by paper and paper boards. Paper boards, paper bags and corrugated boards have become popular forms of packaging for a variety of products from groceries to garments.

Metal containers are excellent for packaging of processed food items, fruits and vegetables, meat products, oil paint etc. However, this type of packing is very costly due to the acute shortage of tin in India. Therefore in recent years, aluminium based packaging has become popular.

With the growth of petrochemical industry a new range of packaging materials has entered to the Indian marketing scene in recent years. Films of low density and high density polythylene (LDPE AND HDPE), metalised polyster film, metalised polyster laminates and polypropylene (BOPP) have become preferred packaging medium for several products. Moreover, plastics are now dominating the packaging field in India. Further more, one of the latest among the various innovation in the packaging field is the tetrapack bticks or aseptic packaging. It is the new development in food packaging. The special feature of this type of packaging is that the package as well as the contents are sterilised and human handling is dispensed with. They are used in pack fruit juices and fruit drinks. Tetrapacks have an edge over cans since their contents have a shelf-life of three months without the addition of preservatives. Moreover, flexible containers are also in use nowadays to pack consumer goods but also in the bulk transportation of commodities.

Legal Standards of Packaging

The marketing of goods especially of consumable items has undergone tremendous innovations over the years. This is in step with the rapid modernisation taking place in trade and industry all over the world. Already in the west, commodities including food stuffs are made available in a ready to use condition, off the shelf in packages. The trend towards marketing the commodities in a packaged form in continuing and large number of items dealt in trade channels are being made available presently in a pre-packed condition. The protection of consumers interest vis-à-vis the sale and distribution of goods in pre-packed form calls for special regulation to be introduced at various levels of trade and industry. When a commodity is sold in pre-packed form it is not easy for the consumers to know at the time of purchasing anything about the contents of the package. It is also not possible for him to ascertain the quality of the goods contained inside the package. The economies of the world have been conscious to the need to regulate suitably the trade in commodities in packaged form.

The rules relating to packaged commodities initially required only the mention of the net contents on the packages. But gradually need for a detailed legislation was felt necessary to regulate their trade. So an expert committee was formed to review the weight and measures Act which suggested elaborate provisions in respect of packaged commodities. These provisions were found to be so important that they were promulgated in the year 1975 under the Defence and Internal Security of India rules, in the form of Packaged Commodities (Regulation) Order 1975. Although prior to this order, drugs sold in packaged form were being marketed with indication as to price since 1962 (under the Drugs Display and Prices Order, 1962 issued under the Defence of India rules, immediately after the Chinese invasion) other articles of mass consumption sold in pre-packed form were not subject to any regulation as regards their price till the issue of the Packaged Commodities (Regulation) Order in 1975. But after the revocation of emergency, the Packaged Commodities (Regulation) Order 1975 ceased to have legal form on 1977. The Government however did not want to terminate the provision of the said order and therefore elaborate provisions were made in the form of the standards of Weights and Measures (Packaged Commodities) Rules 1977.

The rules contain detailed provisions regarding declarations to be made on every packages, the manner in which the declarations are to be made, examinations of packages, export and import of packaged commodities etc.

According to the rules pre-packed commodity means, a commodity which without the purchaser being present is placed in a packaged of whatever nature so that the quantity of the product contained therein has a pre-determined value and such value cannot be altered without the packaged or its lid or cap, as the case may be being opened or undergoing a perceptible modification and the expression 'package' whatever it occurs shall be construed as a package containing a pre-packed commodity.

Therefore, as per the provisions of the definition, the processed fruit and vegetable products fall under the category of packaged commodities.

Rule 6 of the Packaged Commodities Rules, 1976 states the declarations to be made in every package including those containing processed fruit and vegetable products. They are as follows:

1. Name and address of the manufacturer or where the manufacturer is not the packer, of the packer or with the written consent of the manufacturer, of the manufacturer.

2. The common or generic name of the commodity contained in the package. Generic name means the name of the genus of the commodity.

3. The net quantity, in terms of the standard unit of weight and measures of the commodity contained in the package or where the commodity is packed or sold by number, the number of the commodity contained in the package.

4. The month and year in which the commodity is manufactured or pre-packed.

5. The retail sale price of the product.

6. Where the size of the commodity contained in the package are relevant, the dimensions of the commodity contained in the package and if the dimensions of different pieces are different, the dimensions of each such different piece.

7. Such other matters are as specified in these rules. The proviso to Rule 6 (1) states that no declaration as to month and year in which the commodity is manufactured or pre-packed shall be required to be made on.

   (a) Any bottle containing liquid milk, liquid, beverages containing milk as an ingredient soft drink, ready to serve beverages or the like which is returnable by the consumer for being refilled.

(b) Any package containing bread and any uncanned package of *(i)* vegetables, *(ii)* Fruits, *(iii)* ice-cream, *(iv)* butter** *(v)**, *(vi)***, *(vii)* meat, or *(viii)* any other like commodity.**

(i) liquid milk in pouches.

(c) Any package containing metallic product.

(d) Any cylinder containing liquified petroleum gas or any other gas.

(e) Any package containing chemical fertilizer. The month and year in which the commodity is pre-packed may be expressed either in words or by numerals indicating the month and the year or by both.

Also, no declaration as to the sale price shall be required to be made on:

1. Any uncanned package of vegetables, fruits, ice-cream, butter, fish, meat or any other like commodity.

2. Any bottle containing liquid milk, liquid beverages containing milk as an ingredient which is returnable by the consumer for being refilled.

3. Any bottle containing alcoholic beverages or spirituous liquor.

4. Any package containing animal feed exceeding 15 kg or 15 litre.

5. Any package containing a commodity for which a controlled price has been fixed under any law, and

6. Package containing printing ink.

Packaging of the Processed Fruit and Vegetable Products of Assam

The processed fruit and vegetable products of Assam are sold in a pre-packed form. They are packed in both rigid (bottles) and

---

* The word 'Cheese' appearing against *(v)* has been omitted vide amendment to the rules on 17-1-1992.

** The words 'butter' and 'any other like commodity' have been omitted w.e.f. 7-3-1996. w.e.f. amemdment dated 7-12-1995.

flexible (ploythene bags) packages. However, it is often seen that the products sold in rigid packages have received a bright response from the consumers. Moreover, in this form of packages the products remain in a good condition for a pretty long time.

But one of the most important problems faced by the local fruit and vegetable based units of Assam is the procurement of the containers for packing the products.

The containers that is the bottles used for packing the finished products are not easily available within the state. Although polythene bags can be used for packing the items yet such type of packaging stands inferior when compared to the packing of the products in bottles. As a result people do not like to buy the products packed in polythene bags. Moreover, there is a chance of the packet being spoiled by insects and ants if not taken proper care. Further in this kind of packing there is always a chance of the label being spoiled especially in case of pickles which are usually dipped in oil.

These containers are usually brought from places outside the state mostly from Kolkata. While doing so, the expenses of procuring the packaging material becomes very high and as a result it has an impact on the price of the finished products. Although, sometimes containers are found in the local markets but their prices are very high and therefore could not be afforded by the local entrepreneurs. Furthermore, the quantity of the packing materials available are not in proportionate to the quantity required.

Due to such difficulties it is often seen that the entrepreneurs, with the limited capital could not afford to purchase new containers either from the local market at an exorbitantly high price or from outside the state. As a result, such entrepreneurs often buy bottles for packing their finished products from the scrap material vendors and use them after sterilising. Therefore the local fruit and vegetable units face great difficulty regarding the procurement of packaging material. As a result, the cost of packaging the finished products becomes very high and as such the prices of the local products could not compete with the prices of the products from outside the state.

Labelling

A label may be a small slip or a piece of paper placed on, or a printed statement on the merchandise (goods) or its package describing the nature of the product, the content of the package or indicating the destination, ownership, origin or piece. Label is a part of a product. It gives verbal information about the product and the seller. The purpose of labelling is to provide information to the consumer about the product and its uses. Labelling has got social significance too. Consumer criticism centres round charge of false, misleading and defective packaging and labelling. In U.S.A. there is a special Act, Fair packaging and labelling Act (1967) to ensure truth in packaging and labelling to provide consumer protection. But legislation alone cannot do the job unless consumerism consumer pressure and business ethics go a long way in safeguarding innocent and ignorant poor consumers in many countries particularly in India.

Labels in products provide many useful information about the products. These includes the brand name, the name and address of the producer, weight, measure, count, ingredients by percentages where possible. Directions for the proper use of the products, cautionary measures concerning the product and its use, special care of the product if necessary, recipes on food products, nutritional guidelines, date of packing and date of expiry, retail price and unit price for comparison. Since 1991, the rule of mentioning the maximum retail price on the labelling has also been made compulsory in order to avoid the charges of different rate of taxes charged by different manufactures and thereby exploiting the consumers. Labels on food products and drugs contain factual information on which consumers can rely.

Any legal restriction imposed on the packaging of the product is evidenced through labelling, packaged commodities (regulation) order 1975 makes it obligatory on the part of manufactures to show details about the identity of the commodity, its weight, date of manufacture etc. The provision of the enactment is carried out with the help of labelling.

Labelling of Processed Fruit and Vegetable Product of Assam

So far as the processed fruit and vegetable products of Assam are concerned the labelling of such products are not up to the mark. Though labelling is done by the units through offset printing within the state of Assam yet the quality of such labels printed through offset printing in Assam is not adequate in terms of quality that is their colour combination and other presentation. As such these labels are inferior compared to those printed outside the state. This is shown in Table 7.1.

Table 7.1: Opinion of the Entrepreneurs in Percentage Regarding the Quality of Labels Used in the Containers of the Local Fruit and Vegetable Processed Products Printed within the State of Assam

Types of Queries	Opinion of Entrepreneurs	
	Adequate	Inadequate
Opinion of entrepreneurs regarding the quality of the labels printed within the state as compared to outside	10%	90%

Source: Field study.

According to the Table, 90 per cent of the entrepreneurs survey are of the opinion that the quality of such labels are inadequate while 10 per cent of them opined that these are adequate.

Therefore in order to complete with the outside products the labels are to be printed outside the state which increases the cost of labelling. As a result, it has its impact on the price of the product. Again many a times, the entrepreneurs prefer ordinary label than an offset one due to lack of finance.

Therefore the fruits and vegetable based units of Assam face great difficulties in packaging and labelling their products. Many a times the units could not produce the products in bulk due to the non availability of packing material and labels.

Advertising

The dictionary meaning of the term advertising is to give public notice or to announce publicly. Advertising may be defined as the process of buying sponsor identified media space or time in order to promote a product or an idea. The American Marketing Association has defined advertising as follows:

'Any paid form of non personal presentation or promotion of ideas, goods or services, by an identified sponsor.'

Importance of Advertising

Advertising has become increasingly important to business enterprises both large and small. Advertising helps to promote additional sales. It would be difficult for a firm to survive if it does not attempt to promote its product in some manner or another. Not only the business enterprises the non-business enterprises too have also recognised the importance of advertising.

Advertising is equally important both for new and old established product. If an old and established product is not advertised, it will gradually fade away from the notice of the public. Moreover, it acts as a potent instrument in the hands of the new entrant in a competitive market dominated by large and long established firms.

Advertising has offered to society benefits not otherwise available. The criticism that advertising costs too much views an individual expenses item in isolation. It fails to consider the possible effort of advertising on other categories of expenditure. Advertising strategies that increase the number of units sold stimulate economies in the production process too.

Ethics and the Promotion Process

The promotion area is accused of more unethical behaviour than any other marketing area. Ethics deal with what is right and wrong with moral duties and obligations and with a set of moral principles.

Unethical promotion practices include such things as fraud, creating erroneous impression, use of phony testimonials, misleading, brand names and labels. Permanent objective and ethical standards are not available for use by most promotion managers. Each manager determines a standard of promotion ethics based upon moral training in our society, a sense of what is right or wrong, competitive conditions in a particular industry and perhaps some guidance by top management and professional groups. Ethical standards become lower as competition increases. The more competitive an industry, the greater the number of ethical problem encountered.

A great deal of progress has been made since the turn of the century in promotion ethics. Many federal state and local laws are passed by the government in order to improve the promotion ethics.

Advertising in the Fruit and Vegetable Processing Industry of Assam

The fruit and vegetable based units of Assam are mainly run on sole proprietorship basis with limited capital. The Government too have not provided much incentive in regard to the capital contribution to this industry. Therefore lack of finance is one of the root causes for non advertisement of the products of such units. The products of the local fruit and vegetable based units are hardly advertised through any media. This is shown in Table 7.2.

Table 7.2: Opinion of the Entrepreneurs in Percentage Regarding the Advertising of the Fruits and Vegetable Based Products Manufactured Locally

Types of Queries	Opinion of Entrepreneurs	
	Advertised	Not Advertised
Whether the local fruit and vegetable processed products are advertised or not by the entrepreneurs	20%	80%

Source: Field study.

From the above table it is seen that majority of the entrepreneurs of the local fruit and vegetable processed products do not advertise their products, only a few have recourse to advertising of their products. As a result even the local people are quite unaware of the products. This leads to low sales in comparison to the products from outside the state which are widely advertised. Even the retailers hesitate to keep the local products for sale due to their lack of advertisement.

Moreover, due to the lack of business knowledge or managerial crisis, the entrepreneurs do not pay much interest on the advertisement of the products. Furthermore, the scale of operation of such units are very small covering only the customers from the neighbourhood market. As a result, the idea of advertising covering a vast market is still a dream for the entrepreneurs. However, the trade fairs and exhibition help much in this regard.

The products exhibited in these fairs and exhibitions attract people and they come to know about the existence of the product which indirectly advertises the products. In order to boost up the sales of the products the same should be advertised in the local newspapers, magazines, journals, television and, radio. Moreover, free samples of the products should be distributed to make the customers aware about the local products.

Warehousing

Warehousing means the storing of goods in order to create time, place and form utility. It has two distinct and equally important part. Firstly, the physical job of creating and running the network of storage points and secondly the managerial task of controlling the inventory levels.

Every company has to store its goods which they want to be sold. A storage function is necessary because production and consumption cycle rarely matches. Many agricultural commodities are produced seasonally whereas demand is continuous. The storage function overcomes discrepancies in desired quantities and timing.

Storing of both raw materials and finished products are equally important. In case of fruit and vegetable processed products the storing of the finished products and the raw materials are very important because the raw materials are seasonal and perishable in nature while the demand for the finished products remain throughout the year. Therefore cold storage is one of the essential requirements of this industry. Moreover, this industry demands quite a huge premises, for the production and storage to be carried on smoothly. However, the public warehouses can also be used to store the products. In such cases the cost of warehousing goes up.

Warehousing System of the Fruit and Vegetable Processing Units of Assam

The fruit and vegetable processing units of Assam are started by the entrepreneurs mostly in their own houses as household units. Therefore, the space available for production and storing facilities are quite insufficient which causes great inconveniences to these units. This is shown in Table 7.3.

As per Table 7.3 majority of the entrepreneurs are of the opinion that the space available for storing of the finished products are insufficient, only a few are of the opinion that such are sufficient enough.

Table 7.3: Opinion of Entrepreneurs in Percentage Regarding the Space Available for Storing of the Finished Products of the Fruit and Vegetable Processing Units

Types of Queries	Opinion of Entrepreneurs	
	Sufficient	Insufficient
Opinion of entrepreneurs regarding the space available for storing of the finished products of the fruit and vegetable processing units	37%	63%

Source: Field study.

Moreover, there is no cold storage facilities available. The units also cannot recourse to hire space in public warehouses due to lack of finance. Furthermore, due to the non availability of space the finished products could not be stored in large quantities and as a result the supply of the products are not proportionate to the needs of demand. Therefore the area of the market is small and confined only to that locality in which the unit is located. Though very negligible number of units have made an effort to send their products outside the state yet they are not much successful in this regard. Moreover, there is also difficulty in preserving the fruits and vegetables to check their wastage and decay due to the lack of cold storage facilities. This is also proved by the hypothesis No. 1 which runs as a follows:

That the preservation facilities for fruits and vegetables are not adequate to check wastage and decay.

As the result of the survey done on the various local fruit and vegetable processing units it is found that none of the units surveyed throughout the state has cold storage facility to preserve the fruit and vegetables on their own. Therefore most of the units could not make bulk purchases of the raw materials required. The fruit and vegetables being perishable can hardly be stored for 2 to 3 days without cold storage. These units could not afford cold storage because they are mostly started by the entrepreneurs in their own houses as household units and therefore on one hand there is lack of finance and on the other there is also lack of space. So though these are production units their inventory level is nil. This hampers their production as the raw materials are seasonal and perishable in nature. Therefore there is no adequate preservation facilities of the fruits and vegetables to check wastage and decay. However, a negligible number of units have hired cold storage sapce in the public and private sector warehouses which is shown in Table 7.4 but the rate of such spaces are extravagantly high.

Table7.4: Opinion of the Entrepreneurs in percentage regarding the availability of Cold Storage Facility

Types of Queries	Opinion of Entrepreneurs	
	Yes	No
Availability of cold storage facility	18%	82%

Source: Field study.

Out of the total entrepreneurs surveyed which stands to 43, majority of them that is 82 per cent are of the opinion that there is no cold storage facility available. Only a few that is 18 per cent have a positive attitude regarding the availability of cold storage facility.

Regarding the total capacity of cold storage required to preserve the fruits and vegetables as per the wholesale merchants dealing in the free trade there is no prescribed fix size of space required for the preservation of fruits and vegetables. This depends upon the demand and supply of the materials. The fruits and vegetables are hardly preserved. They are sold off to the retailers soon after their delivery. Some particular fruit like apples and grapes are stored and the cold storage space for such are enough which are kept in a private sector cold storage named Chitra cold storage situated at Changsari, Guwahati. Only 18 per cent entrepreneurs regarding availability of cold storage facility are satisfied.

Transportation

Transportation is the physical means of moving goods from one place to another. It plays an important role in the economic development of a nation. Rapid industrialisation cannot take place unless sufficient facilities for transportation is available. Transportation helps in creating time, place and utility. It is with the help of various means of transport that raw materials are transported from the place of their production to the industrial centres where they are converted into finished goods demanded by the customers.

It is again transportation which facilitates movement of goods from producers to users.

Importance of Transport in Marketing

Transport plays an important function in marketing. It consists of all pervasive activities which includes handling, hauling, warehousing, inventory control, physical transportation and delivery. The entire work of assembling and dispersing of goods is done with the help of some form of transport. Transportation today constitutes an important managerial activity primarily due to the fact that it is one of the most costly elements of distribution. Due to this reason that transportation has become the main target of cost cutting elements of distribution. With the improvement in speed, transit time in transport has been shortened to a great extent there by increasing the turnover of capital and products of the business and preventing the risk of price charges. Therefore in a vastly expanded market with cost conscious customers, demanding more value, better service and better transportation management has assumed paramount importance. It is only through better transportation management that the twin objective of quick delivery of merchandise at a relatively cheaper cost could be effectively met.

Transport imparts place utility of goods by moving them from different centres of production to the places of consumption. Goods are now produced thousand of miles away from places where the consumer resides. Nevertheless, a marvellous transport system has ensured a steady flow of goods to the consumer within his easy access. Not only does it gives place utility but it also renders time utility in various ways. With the improvement in speed, transit time in transport has been shortened to a great extent, thereby increasing the turnover of capital and products of the business and preventing the risk of price changes.

Transportation Problems of the Fruit and Vegetable Units of Assam

Assam, a prominent state in the North eastern India is backed by many transportation hurdles. It is cut off from the rest of India during the rainy seasons due to the floods. Not only this, even the linkage between the upper and the lower Assam is very often cut off during floods. As a result transportation within the state also becomes difficult.

The raw materials of the fruit and vegetable processing industry are mostly perishable and seasonal in nature and as such when the raw materials have to cover a long distance from the place of production to the place of processing refrigerated vans are the only useful carrier. But such vans are seldom possessed by the entrepreneurs due to their financial hardships. Therefore most of the raw materials get wasted and decayed till they reach the place of processing. Further the entrepreneurs are not financially so sound enough to hire the services of refrigerated vans. Therefore, the raw materials could not be brought from distant places and the products are to be manufactured from those raw material locally available. Moreover, the cost of transportation of the finished products are also very high. Furthermore, during the rainy season often the upper and the lower Assam is cut off which causes great difficulty in the transportation of raw materials and the finished goods. Therefore due to these transportation problems the entrepreneurs are unable to market their products to the distant places.

Export Marketing

Export marketing is a part of the broad marketing system. Export marketing management involves the marketing not only to other countries but also to some extent within the foreign countries. To a great extent repetitive sales in export marketing are dependent upon how best the marketing is managed in foreign countries. International or multinational marketing refers to the marketing of products and services in more than one nation. It is the performance of business activities that direct the flow of goods and services to consumers or users in more than one nation.

To enter into the Global market a firm has to face many problems. These problems include huge foreign indebtedness, unstable governments, exchange instability, corruption, high cost of product and communication adaptation, tariff and non tariff barriers etc. Inspite of many constraints several companies like to go in for international marketing. There are some strong reasons for it. The first and foremost reason is to earn profit. No firm would do anything to incur losses continuously. The other vital reasons for entering into the global market are, firstly, to make fuller utilisation of capacity. One of the main reasons for a firm to export its products is to make the optimum utilisation of its physical facilities if it does not have a wide domestic market. Secondly, many firms even after having a large domestic market would go for the international market because they find the foreign market attractive enough to make profits. Thirdly, many firms may find it worthwhile to export their products which though in the declining stage of their life cycle in their own countries are comparatively new for foreign countries. Fourthly, exports help in developing domestic trade. The products sold to in the abroad markets also receive good response among the domestic consumers too. Fifthly, the products have a bigger market area and as such the exporter is less dependent upon taste and preference of one particular country. Moreover, due to bulk selling, economical manufacturing is possible. Sixthly, the payment is guaranteed and quick. Moreover, in case of export marketing the exporter is not much affected by the business cycles and political instability of a particular country. If one country has a recession with difficult sales the exporter can export to other market where sales are prosperous.

Export Profile of the Indian Fruit and Vegetable Processing Industry

'India produces the widest variety of fruits and vegetables and is the second largest producer in the world accounting for 11 per cent of the world's production of vegetables and 7 per cent fruits.'[65] But less than one per cent of this production is commercially processed. India's share in the world trade of processed fruits and vegetables is presently less than one percent. 'However, it is readily increasing from 0.275 million tonnes in 1960 to over 0.96 million tonnes in 1998.'[66] The industry in India is highly decentralised as a large number of units are in a cottage and the small scale sectors.

The market for processed food both domestic and global is quite large and its growing. The international market provides substantial opportunity for development of food processing industry. Though our export of processed food is small *i.e.* our share in the global market is around 1 per cent, however it is growing. 'The export of

[65]   Dr. Kurade Naik G. Anand and Dr. Kurade A. Sangarn, Food Processing Industry–Industrial and Export Opportunities; Beverage and Food World, Dairy Management Consult Ant. Sept.–Oct. 1999, pp. 16.

[66]   *Ibid.*

processed food have increased from Rs. 7.8 billion in 1988–89 to over Rs. 25 billion in 1997–98.'[67] 'During the period, the value of processed fruits and vegetables went up from Rs. 640 million to Rs. 1,960 million.'[68] At present the processed fruit and vegetables products are exported to the middle east countries. Price, quality, packaging and delivery schedules are the important factors to be taken into consideration for increasing market penetration abroad. 'The current international trade in processed fruits and vegetables alone is around US$ 6905 million or Rs. 23,477 billion.'[69] 'Out of this processed fruits amount for US$ 4910 million or Rs. 14,246 billion consisting largely of fruits and vegetable juices, juice concentrates, nectares, canned pineapples, semi processed juices, pulps and berries and canned and dehydrated vegetables.'[70]

Because of diverse agro climate conditions varying from tropical to temperate conditions there is a good production base all the year round. With a large consumer base and fairly cheap labour a good potential exists for food processing industry in our country today. Besides liberalised polices of the government helps to export these processed foods in a big way.

Export Marketing by the Fruit and Vegetable Processing Units of Assam

The fruit and vegetable processing industry of Assam is still in its infant stage. The units are started mostly by the sole entrepreneurs in their own houses as house hold units with their own limited capital as such the units are mostly unorganised in nature and could not cover a huge market. Moreover, due to the various difficulties faced by the units such as lack of cold storage facilities, lack of finance, managerial crisis among the entrepreneurs there is no enthusiasm seen among the units to export their products. Further most of the units do not have the FPO licence required for exporting the products. Besides this, in order to export the products the units should also

---

[67] Dr. Kurade Naik G. Anand and Dr. Kurade A. Sangam, Food Processing Industry–Industrial and Export Opportunities; Beverage and Food World, Dairy Management Consult Ant. Sept.–Oct. 1999, p. 18.

[68] *Ibid.*

[69] *Ibid.*

[70] *Ibid.*

have ISO 9000 registration which demands a complicated procedure which the local entrepreneurs find hard to fulfil.

Moreover in order to export the products certain special care should be taken during the process of processing the fruits and vegetables like using of certain special type of cans, maintaining of proper hygienic conditions, special type of packaging etc which the local entrepreneurs hardly fulfil. Furthermore, in case of canned fruits since cutting work is mainly done by the workers manually as such the sizes are not uniform and therefore not accepted by other countries when exported. Further the sugar content in the canned fruits are also not uniform.

The government assistance to boost up the exports of the local products are also not very encouraging. Though a negligible number of firms have exported their products on a sample basis to some neighbouring countries yet they have not received any feedback information regarding the acceptance of their products.

Since most of the entrepreneurs are women carrying out the business as a part of their household activity in their own houses therefore their market is concentrated mostly to their neighbourhood. The production is carried on in small scale and therefore the greater part of the domestic market is still to be captured. So for them entering into the export market is still miles to go. Therefore, they hardly go for exporting their products to other country. This is shown in Table 7.5.

Table 7.5: Opinion of Entrepreneurs in Percentage Regarding the Export of Local Fruit and Vegetable Processed Products to Other Countries

Types of Queries	Opinion of Entrepreneurs	
	Yes	No
Opinion of entrepreneurs whether the fruit and vegetable processed products manufactured by them exported to other countries	27%	73%

Source: Field study.

As per the above table only 27 per cent of the entrepreneurs surveyed have exported the products to other countries. While majority of them that is 73 per cent have not yet done any work for exporting their products.

In order to encourage the export of the local fruit and vegetable processing units cold storage facilities should be provided near to the airports. Moreover, the conversion the Gopinath Bordoloi airport, Borjhar, Guwahati into an international one would provide an incentive to the entrepreneurs to directly export the finished products from Assam to the foreign countries. Furthermore, the state government through its various agencies should take measures to increase the exports of the products. The entrepreneurs should be given national level training to acquire knowledge relating to exporting of the products.

Marketing Research

According to the American Marketing Association, 'Marketing research is the systematic gathering, recording and analyzing of data about problems relating to the marketing of goods and services'. It is the collection and interpretation of facts that help marketing management to get products more efficiently into the hands of the consumers.

Importance of Marketing Research

Marketing research facilitates the managerial decision making process for all aspects of the firm's marketing mix, pricing, promotion, distribution and product decisions. By providing the necessary information upon which to base decisions, marketing research can reduce the uncertainity of a decision and thereby reduce the risk of making the wrong decision. Information is a source that contributes to effective marketing management. It may be used to identify opportunities for enriching marketing efforts. Marketing research can also be used to obtain detail information about specific mistakes or failures regarding managerial judgements. It can also be used to predict or forecast the future conditions of the market. Besides this, it can be used to describe the conditions of the market place.

Types of Marketing Research

Marketing research projects are undertaken to fulfil a wide variety of purposes. This projects vary according to the needs of the various organisation. Depending upon the various types of needs the marketing research projects can be product research, promotion research, distribution research, pricing research, marketing programme research.

Product research is carried on to bring about changes, innovation and develop a variety of styles and models. It is the task of product marketing research to determine consumer requirements to keep abreast of and to channel product technical research and to reach as close to an optimum blending of the two as possible in the products pro-budget, and drawing up sales compensation plans.

Sales effectiveness research includes analyses of time salesmen spend on each of their assigned activities determining ratios of calls to sales made and of expenses to sales, customer surveys to audit salesmen's performance and establishing sales targets and quotas and comparing them with actual sales.

Distribution channel research is concerned with selection of the channels which are to be employed to distribute their products. While conducting research on distribution channel special attention should be paid to the distribution cost analysis. Distribution channel research also includes location research. Location of units depends upon various variables as quantities of goods shipped to differing areas, transportation cost per unit, cost of warehouse operation per unit, times of delivery and service and production capacities. Solutions may be arrived at by use of appropriate mathematical techniques. Factors such as labour supply, living conditions, the political climate and other important intangible considerations must be assessed separately.

Pricing research seeks to obtain information on the qualities that will be demanded at various prices and the corresponding costs of supplying them. Pricing and the establishing of price policies is one of the more complicated problem areas in marketing management and the one in which there has been a least sophistication employed in making decisions. A vast amount of uncertainity surrounds most pricing decisions. The techniques that have been employed in pricing research include customer surveys, dealer surveys, observational studies, correlation and cross classification analyses and experiments of various types.

Marketing performance research is primarily concerned with the identification of problems rather than their solution. Its basic function is to provide information necessary for management to plan and to guage the overall level of performance of the marketing effort. In the process of appraising performance first step is the establishment of specific objectives called performance standards. Marketing performance research is also concerned with the measurement of actual performances and the forecasting of future performance is considered as the second step in the performance appraisal process. The third step in the appraisal process is the comparison of actual and forecast performance with the standards.

The important types of marketing performance research are market potential, market share, sales analysis and sales forecast. Market potential is the amount of a product or service that can be absorbed by the market during a specified period during optimum conditions of market development. Market share is the ratio of a company's sales to the industry sales or either a forecast on a actual basis.

Sales analysis is the analyzing of sales records by the various classification of interest. Sales forecast are estimates of sales for some given future period. Forecast are usually made for each product line and product as well as for the total company sales.

Marketing Research in the Fruit and Vegetable Processing Units of Assam

The fruit and vegetable processing industry of Assam is yet to be developed as a large scale industry. The entrepreneurs of this industry are the local people with limited capital and also limited knowledge. Therefore they do not have enough money and technical knowledge to conduct extensive research programmes regarding the acceptance of the products among the customers. Moreover, no research has been conducted regarding the fixation of prices of their products. Most of the units fix the prices of their products on market at the very beginning of their launching in the market.

Therefore the entrepreneurs of the fruit and vegetable processing units of Assam are yet to learn about the various technique of marketing research applicable in this industry. For this the central and the state government should conduct seminars and workshops on market research of processed food items where expert should be called from noted institutions of the country. Moreover, delegation of entrepreneurs should be send to the large scale fruit and vegetable processing industries outside the state for taking training on their market research programme. Furthermore, the entrepreneurs can conduct a house to house campaign through travelling salesmen by distribution of free samples of the products and thereby take the reactions of the customers regarding the acceptance of the products.

Channel of Distribution

Distribution means to spread out or disseminate. In the field of marketing, channels of distribution indicate routes or pathways through which goods and services flow or move from producer to consumers.

The distribution channel can be formally defined as the basis of competition. They adopt a competitive pricing policy. Besides this, not much research has been conducted for the selection of the distribution channel to sell the products. The products are sold through retailers in the market or through door-to-door salesman engaged by the entrepreneur. Some units sell their products through their own retail outlet. In the field of advertising too till date no sophisticated research technique has been adopted. Since the units are mostly unorganised the products of these units are hardly advertised.

But however in order to know the consumer taste and preference the units have occasionally taken up market research of their products on a very small scale basis. In this context, occasionally the door to door salesman employed by a certain negligible number of units bring information regarding the consumer preference and acceptance about the products yet the number of such salesman employed are very few to cover a huge market. Moreover certain information are also collected by the entrepreneurs from the retailers selling the products. But no effective research technique has been adopted in this regard. Due to the lack of finance the entrepreneurs could hardly afford much expenses in conducting the market research programmes. Moreover, no test marketing is carried regarding the launching of the new products in the market. As a result, certain products are thrown out of the market a set of marketing institutions participating in the marketing activities involved in the movement or the flow of goods or services from the primary producer to the ultimate consumer.'

Types of Channels

Channels of Distribution may be of two types:

1. Conventional and
2. Vertical.

Conventional channels includes,

Producer – Consumer, which is a direct channel and has no intermediary.

Producer – Retailer – Consumer, which is used in case of speciality goods.

Producer – Wholesaler – Retailer – Consumer, which is the most widely used channel.

Producer – Agent – Wholesaler – Retailer – Consumer which is used mostly by relatively small manufacturer with a limited product line selling products in a widely dispersed market.

In case of industrial goods the following channels are used.

Producer – Industrial user (Direct channel)

Producer – Industrial distributor – User.

Producer – Agents – User

Producer – Agents – Industrial distributor – User.

Vertical integrated marketing channels are of three types.

Corporate system where single firm owns both product and distribution facilities. Contactual system where independent firms are employed on a voluntary basis to develop an efficient distribution channel.

Administrative system where the manufacturer controls the marketing of a particular line of merchandise than a complete store operation.

Channel Choice

A large number of distribution channels are available to the manufacturer for bringing his product to ultimate consumers. Out of the various channels available the manufacturer has to choose one which would best be suited for the distribution of his products. In doing so, various factors should be taken into consideration which includes the nature of the product, its unit value, its technical characteristics, its degree of differentiation from competitive products and other product characteristics, financial resources and available expertise with the company and the availability of suitable middlemen.

Channel of Distribution in the Fruit and Vegetable units of Assam

Most of the fruit and vegetable processing units of Assam are operating on a small scale basis with limited capital and production. The products are mostly sold in the local market. Only a few entrepreneurs send their products to the distant markets. Because of the limited capital, narrow circulation of production and the concentration of the products in the local markets, the entrepreneurs do not go for a very long channel of distribution. Usually the entrepreneurs employ certain persons as their selling agents who travel from shop to shop and deliver the products to the retailers who in turn sell those to the consumers. Again in certain cases it is also found that the entrepreneurs send their agents that is the travelling salesman for direct marketing. In other words, they adopt a door to door selling system. Only a very few units have their own retail outlet where they sell their own products.

The local retailers too play an important role on the sale of the products. During the field study among the retailers it was found that the retailers hesitate to keep the local products in their stores firstly, as the rate of sale of such products are low due to lack of advertisement and also at times due to the poor quality of the same. This is the shown in Table 7.6.

Table 7.6: Opinion of the Retailers in Percentage Regarding as to Whether they deal in Local Fruit and Vegetable Based Products

Types of Queries	Opinion of Entrepreneurs		
	Locally Processed Fruit and Vegetable Based Products	Fruit and Vegetable Based Products Coming from Outside the State	Both
Opinion of retailers as to in which type of fruit and vegetable based products do they deal the most	Nil	63%	36%

Source: Field study.

Out of the 150 retailers surveyed all over Assam, only 36 per cent deal in both the locally processed fruit and vegetable products along with those coming from outside the state, while 63 per cent of them deal exclusively on the fruit and vegetable based products coming from outside the state. There is no retailer dealing exclusively on the local fruit and vegetable based product because the rate of sale of such products are very low compared to the those marketed from outside. Secondly, the local entrepreneurs usually do not give credit on their products to the retailers. At times even if they allow credit the expected period of such credit is only 1 week which is very short compared to the credit facility received from the products marketed from outside the state.

So, inspite of the good margin of profit on the local fruit and vegetable processed products the retailers hesitate to deal in them as their demand is less compared to the products from outside the state.

Chapter 8

Consumer Behaviour Towards Processed Foods: An Empirical Test

Behaviour is a mirror in which every one displays his image

Goethe

Definition of Consumer Behaviour

In the present times the consumer is well recognised as the king. He is the centre around which the marketing management revolves. Therefore it is as much necessary to know who are the people that consume products or use services so as to ascertain why they do so. Though human mind is the most abstruse thing in the World yet the growth of psychology and the behaviourial sciences have made possible to have some sort of X-ray of the human mind. One cannot observe the mind but one can see and scrutinize behaviour.

The chief aim of consumer research is to study the behaviour of consumer in order to find out the principal reasons or factors behind it and then to influence the behaviour with the help of this information.

Consumer behaviour can be defined as, 'All psychological, social and physical behaviour of all potential consumers as they become aware of evaluate, purchase, consume and tell others about products and services.'[71]

'Consumer behaviour refers to those acts of individuals directly involved in obtaining and using economic goods and services including the decision processes that precede and determine these acts (Engel, Blackwell and Kollat, 1978).'[72]

The post liberalisation period has resulted in many companies entering the markets with offering of their goods and services. This has made each marketer to realise that he has to constantly upgrade the consumers knowledge about his products by finding new dimensions. This is because there has been a change in the physical behaviour of the consumer. The consumer of the yesteryears was a silent person who without complaining purchased the goods from the market place. There is a new consumer emerging today. In the present times the consumer is the choice empowered one who will be the decider as which products to buy and which to reject. So, therefore in such a situation it is utmost necessary for the producers and marketers to make a constant study of the consumer behaviour in order to build a strong base of his products in the market.

Determinants of Consumer Behaviour

Consumers vary tremendously in their ages, income, education level, mobility pattern, tastes and preferences. Earlier economics was regarded as marketing's mother discipline. This was because it was seen that consumer purchase decision is governed by their economic and mental forces. Mental forces like fear, pride, fashion, possession, sex or romance, vanity etc. create desires and wants in the minds of the consumers but economic forces such as purchasing power may come in the way of satisfying those wants hence the consumer has to choose between those wants and select the products according to the priority of consumption. However the rapid changes taking place in the external environment has also had a bearing on the consumer

---

[71] Nair, R. Suja. Consumer Behaviour, Himalaya Publishing House, 1999, p. 3.

[72] Banerjee Mrityunjoy. Essentials of Modern Marketing. Oxford and IBH Publishing Company Pvt. Ltd., 1988, pp. 106.

behaviour and is seen to have complicated the manner in which a consumer behaves. Researches in this field have shown that the light has been shed on the human behaviour on in general and the study of behavioural sciences be extended to the behaviour exhibited by individuals in their roles as buyers/consumers. So consumer behaviour can be said to be mending of all those bodies of knowledge concerning with human behaviour-Behavioural sciences. To facilitate better understanding of the inter disciplinary dimension of consumer behaviour, concept has been borrowed from other disciplines having a bearing on consumer behaviour.

Economic Model

Economic model of consumer behaviour is unidimensional. This model lay emphasis on only one aspect of the individual buyers that is income. According to this model, lower the price of the product the bigger the quantity that will be bought. Again the higher the purchasing power the higher the quantity that will be bought. Moreover, the lower the price of a substitute the lower the quantity that will be bought of the original product and the higher promotional expenditure the higher will be the sales. The economic model ignores many other aspects such as perception, motivation, learning, process, attitudes, personality, culture and social class. Therefore a model dealing only with price and income influences on buyer behaviour and ignoring many other individual (psychological, social, cultural) and marketing variables (product variations and innovations, distribution system, marketing communications) cannot be considered adequate in modern customer oriented marketing philosophy.

Psychological Model

Psychology is referred to as a science, a study of minds and its processes (study of minds). Thus psychology as a branch of behavioural sciences tries to understand how an individual's mind works while taking decisions. Here psychology tries to understand the role played by needs and motivation, personality, perception, learning, level of involvement and attitude in influencing the consumer's decision making process.

Motivation

Motivation is the desire to act, to move, to obtain a goal or an objective. It is a mental phenomenon. It is affected by perceptions, attitudes, personality traits and by outside influences such as culture and marketing efforts. Needs may be physiological, social and psychological. Any urge moving or prompting a person a purchase decision is called a buying motive. Motivation research as a part of marketing research tries to answer the 'why' of consumer behaviour. It also contributes to product development and advertising creativity.

Perception

Perception has been defined by social psychologist as the complex process by which people select, organise and interpret sensory stimulation into a meaningful and coherent picture of the world. It determines what is seen and felt by the consumers when numerous stimuli are directed to them everyday by messages broadcast by the marketers through their promotional devices. Motivation provides a basic influence upon buyer behaviour while perception is operationally critical. It causes the behaviour in a certain way.

Learning

Learning is the central topic in the study of human behaviour that result from previous experience and behaviour in similar situations. Learning is the product of reasoning, thinking information processing and of course perception. Buyer behaviour is critically affected by the learning experiences of buyers.

Attitude and Beliefs

A belief is a description thought that a person holds about something. Attitude is defined as an emotionalised predisposition (inclination) to respond positively or negatively to an object or class of objects. The concept of predisposition includes our familiar concepts of attitudes, beliefs, goals and values. Attitudes affect both perception and behaviour. In general an attitude is a state of mind or feeling. It induces a predisposition to behave in some way. It is very important in explaining buyer behaviour. Attitudes eventually influence buying decision which people make and therefore marketers are deeply interested in buyer's attitudes, beliefs, values and goals *i.e.* buyers predisposition. Attitude research offers a useful device for explaining and predicting buyer behaviour. Knowledge of consumer attitudes can provide a good basis for improving products, re-designing packages and developing and evaluating promotional programmes. Consumers resist a change in their attitudes. But a change in the attitude leads to a change in buying behaviour. Promotional devices are essential to change purchasing attitudes and modify buyer behaviour.

Personality

Personality is a complex psychological concept. Its primary features are self concept roles and levels of consciousness. The self concept refers to how a person sees himself and how he believes others to see him at a particular time. Each individual plays many different roles-father, mother, wife, friend, co-worker executive etc. The consumer behaviour is influenced by the particular role upon which a buyer is concentrating at a given time.

Social and Cultural Influence

Family

Most consumers belong to a family group. The family can exert considerable influence in shaping the pattern of consumption and indicating the decision making roles. Personal values, attitudes and buying habits have been shaped by family influence. The members of the family play different roles such as influencer, decider, purchaser and user in the buying process. Marketing policies regarding product promotion and channels of distribution are influenced by the family members making actual purchases.

Reference Group

Buyer behaviour is influenced by the small groups to which the buyer belongs. Reference group are the social economic or professional group and a buyer uses to evaluate his or her opinions and beliefs. Buyer can get advice or guidance in his or her over thoughts and actions from such small groups. Consumers accept information provided by their peer groups on the quality of a product, on its performance, style etc which is hard to evaluate objectively. Group members provide relevant and additional information which cannot be provided by mass media. A satisfied customer becomes the salesman of the product.

Social Class

Consumer buying behaviour is determined by the social class to which they aspire rather than by their income alone. Social classes may act as one criterion for market segmentation.

Culture

Culture, represents an overall heritage, a distinctive form of environmental adaptation by a whole society of people. It includes a set of learned beliefs, values, attitudes, morals, customs, habits and forms of behaviour that are shared by a society and transmitted from generation to generation within that society. Culture influence is a force shaping both patterns of decision making from infancy. Our cultural institutions provide guidelines to marketers. Technological changes, education and travel have considerable influence on culture. The behaviour of the consumers vary according to the culture to which they belong.

Consumer Behaviour in the Indian Society

Understanding of market and its wide dimension is very important to a marketing decision maker. Consumers differ widely in terms of space, time, perception, value and ownership.

In India, consumer behaviour is governed by many factors. India is a subcontinent in terms of size. The regional and climate variations have a great impact in the clothing and food habits of the Indian consumers. The variations in consumer behaviour also occurs due to the difference in the income level and life of the consumers. India is one of the world's highly populated country. The population pattern of the country also to a great extent influence the behaviour of the consumers. Moreover, consumer behaviour in India to a large extent is influenced by the sex composition. Today the urban housewife is an active partner and plays a major role in the purchase decision of her family along with her husband. Another important factor which influences the behaviour of the Indian consumer is their literacy level. Depending upon the level, of literacy amongst the target consumers the marketers will have to design a suitable communication mix for promoting a particular product or service. Further the income level also has a great impact on the purchase decisions of the consumers. Income level can be divided into two, namely, the disposable income (income minus taxes) and discretionary income *i.e.* income left after paying taxes and meeting expenses related to food, clothing, shelter and the other necessary items. Income credit and assets are objective elements of the purchasing power of Indian consumers. However, economic ability must be combined with the willingness to buy. Moreover purchasing power also depends upon the per capita income of its largest market.

India is a land of various religions. In addition to this, there are various sects, sub-sects, caste and sub-caste. Each religion has its own set of custom rituals and practices which are being followed from generations altogether. There are some religious constraints which influence the consumption and buying behaviour of individuals. Moreover the style of dressing varies among the states and the religious communities of India. Festivals are also an important aspect of Indian culture. The dressing style, food habits and celebration of festivals by the consumers greatly depend on the religious practices followed. Thus the consumption pattern of the Indian consumer is based on the values, beliefs and customs inculcated right from the time of birth.

An understanding of the consumer behaviour is extremely important in the new environment of competition and consumer awareness especially in the food product markets. Regarding the food products the consumer is quite conscious about his health while purchasing it. They are also quite aware or the quality standards like Agmark, ISO 9000, FPO and so on. Moreover, the consumer today are interested in the freebies which in recent years are very often offered in the food products like gifts, discounts etc.

Consumer Behaviour Towards the Fruit and Vegetable Products of Assam

To have an idea about the consumer behaviour towards the fruit and vegetable processed products of Assam an extensive empirical survey among the consumers was conducted throughout the state of Assam by the distributions of questionnaire among the consumers of different income level. Altogether 150 consumers were surveyed all over Assam belonging to different income level.

The income of the consumers were divided into 3 categories *viz.*, below 5000, 5000 to 10,000 and above 10,000. 50 consumers belonging to each income group were surveyed.

As per Table 8.1 all the people belonging to different income consume processed food and purchase fruit and vegetable processed products from the market. Regarding whether it is a must item in their daily food most of the consumer gave a negative answer. Only a few was of the opinion that it is a must item in their daily food. Out of the 50 consumers belonging to the income level below 5000, 17 of them *i.e.* 34 per cent was of the opinion that it is must item in their daily food while 33 of them *i.e.* 66 per cent had given a negative

Table 8.1: Opinion of the Consumers Towards Processed Food Products
(Number of Consumers Surveyed = 150, Number or consumers in each income level = 50)

Monthly Income of Consumers	Consumption of Processed Food				Purchase of Jam, Jelly, Pickles from the Market				Must Item in your Daily Food				Preference Given to Home Made/Outside Prepared Items			
	Yes		No		Yes		No		Yes		No		Yes		No	
	No.	%	No.	%	No.	%	No.	%	No.	%	No.	%	No.	%	No.	%
Below 5000	50	100%	x	x	50	100%	x	x	17	34%	33	66%	50	100%	x	x
5000–10000	50	100%	x	x	50	100%	x	x	12	24%	38	76%	44	88%	6	12%
Above 10000	50	100%	x	x	50	100%	x	x	21	42%	29	58%	50	100%	x	x

Source: Field survey among the consumers.

Table 8.2: Opinion of the Consumers Towards Processed Food Products
(Number of Consumers Surveyed = 150, Number or consumers in each income level = 50)

Monthly Income of Consumers	Standard of Local Products in Terms of Quality				Preference Given to a Particular Brand				Preference Given to Local/Outside Products				Price of Local Products Compared to Outside Products					
	Adequate		Inadequate		Yes		No		Local		Outside		High		Average		Low	
	No.	%	No.	%	No.	%	No.	%	No.	%	No.	%	No.	%	No.	%	No.	%
Below 5000	33	66%	17	34%	42	84%	8	16%	17	34%	33	66%	33	66%	17	34%	×	×
5000–10000	17	34%	33	66%	28	56%	22	44%	11	22%	39	78%	28	56%	22	44%	×	×
Above 10000	10	20%	40	80%	20	40%	30	60%	15	30%	35	70%	50	100%	×	×	×	×

Source: Field survey among the consumers.

Table 8.3: Opinion of the Consumers Towards Processed Food Products
(Number of Consumers Surveyed = 150, Number or consumers in each income level = 50)

Monthly Income of Consumers	Awareness of Local Fruit and Vegetable Products				Local Products Easily Available				Preference for Change of Taste				Reselling		Use of Used Bottles			
	Yes		No		Yes		No		Yes		No				Home Used		Waste Products	
	No.	%	No.	%	No.	%	No.	%	No.	%	No.	%	No.	%	No.	%	No.	%
Below 5000	33	66%	17	34%	x	x	50	100%	50	100%	x	x	17	34%	33	66%	x	x
5000–10000	17	34%	33	66%	11	22%	39	78%	50	100%	x	x	x	x	17	34%	33	66%
Above 10000	21	42%	29	58%	x	x	50	100%	50	100%	x	x	x	x	40	80%	10	20%

Source: Field survey among the consumers.

response to the particular query. Similarly among the consumers falling under the income level between 5000–1 0000, 12 of them *i.e.* 24 per cent gave a positive answer and 38 of them *i.e.* 76 per cent gave a negative answer. Again in the income level above 10,000, 21 of them *i.e.* 42 per cent gave a positive response and 29 of them *i.e.* 58 per cent gave a negative response.

The same table also shows whether consumers prefer processed home made items in terms of those available in the market. Regarding this all the consumers belonging to the income group below 5,000 and above 10,000 prefer home made items but the opinion of the consumers belonging to the income group between 5,000-10,000 does not tally with the consumers of the other two groups. In this income group out of 50, 44 of them *i.e.*, 88 per cent prefer home made item and 6 of them *i.e.*, 12 per cent prefer market items.

Table 8.2 shows consumer behaviour towards processed food items in terms of their preference to any particular brand of the fruit and vegetable products. In the income level below 5,000, 42 out of the total 50 consumers *i.e.*, 84 per cent are brand conscious and like to stick to a particular brand.On the other hand, 8 out of 50 *i.e.*, 16 per cent do not prefer any particular brand.

In the income group between 5,000–10,000, 28 of the total 50 consumers *i.e.*, 56 per cent prefer a particular brand and 22 of them *i.e.*, 44 per cent do not prefer any particular brand. Again in the income group above 10,000, 20 of the total 50 consumers surveyed *i.e.*, 40 per cent prefer a particular brand and 30 of the total 50 *i.e.*, 60 per cent do not prefer any particular brand.

Along with this, Table 8.2 also shows whether preference is given to local fruit and vegetable processed products or those from outside the state. The information collected from the different income groups shows that out of the 50 consumers surveyed, in the income group below 5,000, 17 of them *i.e.* 34 per cent prefer local products and 33 of them *i.e.* 66 per cent prefer outside products. In the income group between 5,000–10,000, 11 consumers that is 22 per cent out of the total 50 prefer local products and 39 consumers *i.e.* 78 per cent prefer products from outside the state of Assam. Again in the income group above 10,000 15 of them *i.e.*, 30 per cent prefer local products and 35 of them *i.e.*, 70 per cent prefer products processed outside the state. Therefore majority of the consumers prefer outside products. This is also proved by the hypotheses no. 4 which states that the consumer behaviour is not friendly and encouraging to local products (by using x^2 chi-square test).

Hypothesis 4

Table 8.4: Consumer Preference Towards the Local Products Using x^2 (Chi-square Test)

Preference to Local/ Outside Products	Income Level of the Consumers		
	Below 5000	5000–10000	Above 10000
Local products	17	11	15
Outside products	33	39	35

Let us take the hypothesis that consumer behaviour is not friendly to local products.

17	11	15	43
33	39	35	107
50	50	50	150

14	14	15	43
36	36	35	107
50	50	50	150

Expected of (AB)

$E_{11} = (A \times B)/N = (50 \times 43)/150 = 14.3$

$E_{12} = (A \times B)/N = (50 \times 107)/150 = 35.6$

Applying x^2 test

0	E	$(0-E)^2$	$(0-E)^2/E$
17	14	9	0.64
33	36	9	0.25
11	14	9	0.64
39	36	9	0.25
15	15	0	0
35	35	0	0
			1.78

For $r = 2$

$$x^2\,0.05 = 5.99$$

The calculated value of x^2 is smaller than the value of x^2 at 5 per cent level of significance for 2d.f

The null hypothesis is accepted and we can conclude that consumer behaviour is not friendly to local products.

Moreover regarding the price of the local products in comparison with the outside products none of the consumers belonging to either group is of the opinion that the price of the local products are low. Most of the consumers of all the income group are of the opinion that the prices of the local products are high.

In the income group below 5,000, 33 consumers *i.e.*, 66 per cent out of the total 50 surveyed are of the opinion that the prices of the local products are high. Similarly, 28 consumers *i.e.*, 56 per cent belonging level between 5,000–10,000 income level are also of the opinion that the prices of the local products are high.

In the income group above 10,000 all of the 50 consumers surveyed *i.e.*, 100 per cent are of the opinion that the prices of the local products are high. Similarly in the income group below 5,000, 17 consumers *i.e.*, 34 per cent are of the opinion that the local products have an average price. Again in the income group between 5,000–10,000, 22 consumers *i.e.*, 44 per cent are of the opinion that the locally produced products have an average price. Further, the same table also shows the standard of the local products in terms of quality. In this context 33 consumers *i.e.*, 66 per cent in the income level below 5,000 are of the opinion that the quality of the local products are adequate while 17 consumers *i.e.*, 34 per cent opines that the quality of the local products are not adequate.

Again coming to the income group between 5,000–10,000, 17 consumers *i.e.*, 34 per cent are of the opinion that the quality of the products are adequate while 33 consumers out of the total 50 *i.e.*, 66 per cent are of the opinion that the quality of the products are inadequate. In the income level above 10,000, 10 consumers *i.e.*, 20 per cent are of the opinion that the quality of the products are adequate while 40 *i.e.*, 80 per cent are of the opinion that the products are inadequate in terms of quality.. This is shown and proved in the hypothesis no. 3, which states that the local products are not competitive in terms of quality and costs to edge out the products from outside the state (through the field study carried among the consumers by distributing questionnaire).

Hypothesis 3

That the local products are not competitive in terms of quality and costs to edge out the products from outside the state.

Opinion of the consumers regarding the quality and cost of the local fruit and vegetable based products.

Table 8.5: Opinion of the Consumers Regarding the quality of the Products. (In Number and Percentage)

Opinion of the Consumers regarding the Quality of the Local Products	Quality of the Products		
	Adequate	Inadequate	Total
Consumers having income below 5000	33 (66%)	17 (34%)	50
Consumers having income between 5000–10000	17 (34%)	33 (66%)	50
Consumers having income above 10000	10 (20%)	40 (80%)	50
Total	**60 (40%)**	**90 (60%)**	**150**

Source: Field study.

As per the Table 8.5 out of the total 150 consumers surveyed, 60 of them are of the opinion that the quality of the local products are adequate but majority of them *i.e.,* out of 150,90 of the consumers are of the opinion that the quality of the products are not adequate. Therefore 40 per cent of the consumers are of the opinion that the quality of the products are adequate while 60 per cent of them are of the opinion that the quality of the products are not adequate. Hence a conclusion can be drawn that the local products are not competitive in terms of quality to edge out the products from outside the state.

As per Table 8.6 out of the total 150 consumers surveyed most of the consumers *i.e.,* 111 (74 per cent) are of the opinion that the prices of the local products are high while 39 (26 per cent) of them are of the opinion that the prices of the local products are average compared to the outside products. But none of the consumers are of the opinion that the prices of the local products are low compared to the outside products. Therefore we can draw a conclusion that the local products are not competitive in terms of quality to edge out the products from outside the state.

**Table 8.6: Opinion of the Consumers Regarding the Cost
of the Local Products**

Opinion of the Consumers regarding the cost of the the Local Products	Price of the Products			
	High	Average	Low	Total
Consumers having income below 5000	33 (66%)	17 (34%)	X	50
Consumers having income between 5000–10000	28 (56%)	22 (44%)	X	50
Consumers having income above 10000	50 (100%)	X	X	50
Total	**111** (74%)	**39** (26%)	**X**	**150**

Source: Field study.

Regarding the easy availability of the local fruit and vegetable based products Table 8.3 shows that all the consumers surveyed belonging to the income level below 5,000 and above 10,000 are of the opinion that the products are not easily available while 11 out of total 50 consumers *i.e.*, 22 per cent belonging to the income level between 5,000–10,000 are of the opinion that the products are easily available while 39 of them *i.e.*, 78 per cent are of the opinion that the products are not easily available.

Another important thing suggested by all the consumers belonging to all the income groups is a change in the taste of the local products that is the entrepreneurs should try to modify their products and give abetter taste to their products and switch off to the processing of untapped fruits and vegetables. Regarding the use of used bottles the consumers of different groups have different opinion 17 out of 50 consumers that is 34 per cent of the income group below 5,000 in their homes while 33 that is 66 per cent of them throw the used bottles as waste products. 80 per cent of the consumers that is 40 of them in the income group above 10,000 make use of the used bottle in their household work while 20 per cent of them that is 10 consumers throw them as waste products. Regarding the awareness of the existence of any locally produced fruit and vegetable processed products units 33 consumers that is 66 per cent belonging to the income level below 5,000 are aware of the existence of the locally produced fruit and vegetable based products units but 17 consumers that is 34 per cent are not aware of any such products. Again in the income group between 5,000–10,000, 17 consumers that is 34 per cent are aware of the existence of locally produced fruit and vegetable based products units. While 33 of them that is 66 per cent have no idea about any such products. In the income group above 10,000, 21 of them that is 42 per cent are aware of such products while 29 of them that is 58 per cent have no such idea.

Therefore it is found that there is a mixed response on the part of the consumers regarding the acceptance and rejection of the local fruit and vegetable.

Chapter 9
Summary and Findings

In the preceding chapters the various aspects of the fruit and vegetable processing industry of Assam have been discussed in a detailed way laying special emphasis on the marketing aspect of such products. The various discussions include the availability of raw materials (fruits and vegetables) for the processing units in the North east, the prospect of the food processing units of Assam, the various organisational problems of the fruit and vegetable industry, special marketing problems of such units and the behaviour of the local consumer towards processed food. The entire discussion of the whole research work are summarised in this chapter.

Chapter 1

This chapter is the summary of the proposed work undertaken in this exercise. Here efforts have been made to discuss all the relevant topics of the contemplated study in a nutshell. In this context, mention can be made of the agricultural background of the North east particularly Assam, the food processing industry of the North east and Assam and the institutions providing help to the industry.

Moreover the chapter also highlights the significance of the whole work and its relevance to the economic and social development of Assam. Besides this, the chapter also discusses the objective, methodology and limitation of the study and also the various paper works done on this topic.

Chapter 2

As the main emphasis of the whole work is on the marketing aspect of the local fruit and vegetable processing units of Assam. This chapter is a theoretical one which throws light on the conceptual framework of marketing management. Attempts have been made to discuss the basic areas of marketing management from the barter stage to the introduction of money economy. This part of the chapter also stresses on the changes in the marketing system during the industrial revolution which originated during the end of the 18th century. Moreover, it also depicts the various marketing channels adopted by different countries during different times which are much similar to the present day distribution. As a whole, in this part of the chapter, attempts have been made to bring out an overall discussion on the emergence of marketing right from the barter stage to the stage of consumer oriented marketing.

The chapter also throws light on the various definition forwarded by the various authorities of marketing. Moreover, discussions are also made on the significance and benefits of marketing in the highly competitive and changing market condition of the present day economy. Further, the chapter also discusses the five distinct concept of marketing *i.e.* exchange concept, production concept, product concept, sales concept and marketing concept. The chapter also brings into light the modern marketing scenario in the globalised economy. In this context, discussions are made on the role played by the world trade organisation (WTO) in the marketing of products in the present day globalised economy. Further, discussions are also made on the innovations brought about in the field of packaging of the products for the world wide globalised market.

Chapter 3

The main raw material of the fruit and vegetable processing industry comes from agriculture alone. As such this chapter is entirely entrusted in discussing the agricultural output of the north eastern region particularly Assam. The chapter makes a thorough discussion on the various aspects physical features of the North eastern region particularly Assam. The discussion includes the boundaries of the region, the natural division, the climate conditions including the rainfall, soil, land use pattern. Likewise the chapter also discuss the physiographic frame of the state of Assam. The discussions include Location, Boundaries, Natural division, Climate, Rainfall, Soil and Land use pattern.

The chapter also makes a detailed discussion on the agricultural scenario of Assam and also of the entire North eastern region. The discussion includes the major crops of the region along with their year wise production. Further the chapter also discuss the role played by agriculture in the economic development of Assam and the entire North eastern region.

Chapter 4

This chapter makes a detailed discussion on the prospect of the food processing industry of Assam. The chapter also tries to bring a close profile of the food processing industry in the entire state of Assam in a district wise detail analysis. Though there is no specific definition of food processing, yet in this chapter an effort is made to put forward a very clear definition of food processing. The chapter also makes a discussion as to the total number of food processing units in the entire state of Assam making a detailed analysis of the various food processing units in Assam.

Further the chapter also makes a detailed district wise analysis of the various food processing units in each district of Assam with reference to the various categories of the food processing units in each district. The chapter also discuss the future prospect of the food processing industry in Assam.

Chapter 5

In this chapter an effort has been made to make a thorough discussion on the various organisational problems of the fruit and vegetable based units in Assam. In this context, a detailed discussion is made on the characteristics of the fruit and vegetable processing industry with more special emphasis to the fruit and vegetable based units of Assam.

Moreover, discussions on the various organisational problems include production which is subdivided into purchase, storage, quality control, transportation, food technology and FPO requirements; finance, personnel which include labour, wages and training; marketing which comprises of packaging, labelling, advertisement, warehousing, market research, transportation, export marketing and channel of distribution.

Chapter 6

Government agencies play a vital role in the development of any industry. In this chapter a detailed discussion is made about the role played by the government agencies in the development of fruit and vegetable industry. In this context, a detailed discussion is made on the various schemes of the central and the state government under which various assistance are provided to the food processing industry. Moreover, the chapter also discusses the various assistance provided by the government and semi government agencies to the different food processing industry of Assam especially the fruit and vegetable processing ones. The various government and semi government institutions are North Eastern Regional Agricultural Marketing Corporation (NERAMAC), Assam Small Industries Development Corporation (ASIDC), Assam Financial Corporation (AFC), Agricultural and Processed Food Products Export Development Authority (APEDA), Small Industries Service Institute (SISI), District Industries Centre (DIC), National Small Industries Corporation Limited (NSIC), Indian Institute of Entrepreneurship (IIE), Small Industries Development Bank of India (SIDBI) and National Horticulture Board (NHB), North Eastern Industrial and Technical Consultancy Organisation Limited (NEITCO).

Chapter 7

Since the main aim of the whole work is to focus on the marketing aspect therefore this chapter deals entirely with the marketing problems of the fruit and vegetable based units of Assam. The various marketing problems include packaging, labelling, advertising, warehousing, transportation, export marketing, market research and channel of distribution. The fruit and vegetable units of Assam is lagging behind mainly due to the inefficient marketing of the local products.

In this context, the packing and labelling problem occupies the topmost priority because often the entrepreneurs surfer due to the non availability of packing materials and also due to the products. Secondly, the products are not able to capture the market because there is lack of advertisement of such and as a result the consumers are not aware of them. Thirdly this chapter also highlights the warehousing problem faced by the industry. In this context, mention can be made about the non availability of space to preserve the raw materials and the finished products. Moreover there is no cold storage

facility required by this industry. Fourthly, the chapter deals with
the transportation problems faced by this industry which include
the lack of refrigerated vans to transfer the raw materials from distant
places to the place of processing. Moreover the transportation cut off
during the rainy season due to floods and so on. Fifthly the chapter
also deals with the export of the local fruit and vegetable processed
products. In this context it can be mentioned that exporting of this
products on commercial basis to other countries is a far cry. There
are many reasons for it. They are lack of enthusiasm and managerial
crisis among the entrepreneurs, absence of FPO and ISO 9000
registration of the products, lack of technical knowledge of preparing
such products and so on. Sixthly discussions are also made on the
market research of such products. No steps are taken for market
research of such products and as a result the products are unable to
capture the market. Even though at times such effort are made by
certain entrepreneurs yet such are carried in an unorganised way.
Seventhly the chapter also discusses the problems in the distribution
aspect of such products. The entrepreneurs do not recourse to a very
long channel of distribution as there is lack of capital among the
entrepreneurs of such industry and also due to the concentration of
the market of such products in local areas.

Chapter 8

This chapter deals with the attitude of the consumers towards
the local fruit and vegetable based products. In the present era of
modern marketing consumer is the pivotal around which the whole
marketing revolves. Therefore in order to develop the marketing of
any product study of consumer behaviour is very crucial. So keeping
in view this idea efforts are being made in this chapter to highlight
the response of the consumers towards the processed food in an
empirical way by means of questionnaire supplied to them.
Moreover, along with the definition and determinants of consumer
behaviour, efforts are also been made to study the consumer behaviour
in the Indian society at large. Regarding the consumer behaviour
towards the fruit and vegetable processed products of Assam various
information are collected from 150 consumers belonging to different
income levels all over Assam who have given their opinions about
the acceptance and rejection of the local products. The consumers
have given a mixed response regarding the acceptance and rejections
of the local fruit and vegetable processed products.

Chapter 9

This chapter is the summary of the whole work. It highlights the main points of each chapter in a nutshell serially from Chapter 1 to Chapter 10.

Chapter 10

In this chapter suggestions are forwarded for the upliftment of the local fruit and vegetable industry of Assam. This includes the steps to be taken from the field to the processing units. Suggestions are also made in context with marketing of the finished products including the export marketing, market research. These suggestions include the steps to be taken at the individual level by the entrepreneurs and also by the government.

Chapter 10

Strategy for Development of Fruit and Vegetable Based Industry in Assam

Food processing industry fall under the wide category of agro based industries which occupies a very important place in the economy of the country. India is an agricultural state. The Indian agriculture has been the source of raw materials to many leading industries. The food processing is one of them. In the recent years the importance of this industry is being increasingly recognised both for the generation of income and for generation of employment. The employment generated by this industry is approximately 1.5 million, which constitute 19 per cent of the country's industrial labour. The industry also contributes 19 per cent of the total number of manufacturing units, 13.5 per cent of the total output 6.5 per cent of the value added. Moreover food industry accounts for only 5.2 per cent of the total investment in the industry in the whole of the country yet contributing 18 per cent GDP.

The food processing industry cover a wide spectrum of products, of which the fruits and vegetables based products cover an unique position. This industry plays a vital role in the effective utilisation of horticultural products. Besides, reducing wastage and losses the fruit and vegetable based units helps in raising rural income by generating employment opportunities.

The region of north east India including Assam is endowed with a wide range of horticultural products due to its unique geographical and climatic condition. Yet the growth and development of the fruit and vegetable based units are not proportionate to the supply of raw materials. As such an attempt is made in this paper to focus the various problems of the fruit and vegetable based units or Assam in general on the basis of an extensive personal survey of entrepreneurs.

Agricultural Scenario of Assam

Assam which is one of the prominent states of North East Region is mainly an agricultural state. The main occupation of the rural people who constitute 90 per cent of the total population of the state is agriculture and more than half of Assam's national income comes from agriculture alone.

The state of Assam has 26,95,941 hectares of agricultural land or 34.3 per cent of the land are available for production of food grains and other crops. Though the average yield of the different crops in Assam compare favourably with all India averages, but considering the excellent climatic conditions, abundant rainfall and fertile soil of the region the yield per acre is quite low. Table 10.1 shows the production of principal crop in Assam.

Table 10.1: Production of Principal Crops in Assam
000'tonnes

Crops	1991–92	1992–93	1993–94	1994–95
Cereals				
Autumn rice	494	614	587	619
Winter rice	2487	2442	2556	2447
Summer rice	216	243	219	213
Maize	11	13	12	13
Wheat	111	79	101	104
Others cereals and small millets	6	5	5	5

Contd...

Table 10.1–Contd...

Crops	1991–92	1992–93	1993–94	1994–95
Pulses				
Gram	2	2	1	1
Tur (Aahar)	5	4	4	4
Rabi	47	45	51	54
Total Foodgrains	**3380**	**3402**	**3536**	**3490**
Sesamum	8	7	7	8
Rape and mustard	178	138	132	150
Linseed	4	4	4	5
Castor	1	1	–	1
Coconut (million metric tonnes)	94	–	–	–
Total oil seeds (excludes coconut)	**191**	**150**	**144**	**164**
Miscellaneous				
Sugarcane	1453	1548	1374	1505
Tobacco	1	1	1	1
Potato	473	388	507	567
Chillies	8	8	8	9
Sweet potato	29	29	30	31

Source: Directorate of Economic and Statistics, Assam, 1995.

Moreover horticulture bears a bright future in Assam and it has every opportunity to be developed here as valuable processed food product. Assam has an abundance of fruits and vegetables which serve as raw materials for the fruit and vegetable based units of this state. Table 10.2 shows the area and fruit crops under production in the state.

Fruit and Vegetable Based Units of Assam

The fruit and vegetable based units of Assam is still in its infant stage. Even though the state has tremendous treasure of horticultural products yet the growth of this industry is to commensurate to the raw materials available for this industry in the state. Table 10.3 shows the number of fruit and vegetable based units in the various district of Assam as on 31–03–98.

Table 10.2: Fruit Crops of Assam in Year 1991–92
Area in hectares, Production in metric tonnes

Fruit Crops		Assam
Pineapple	A	12129
	P	1177594
Orange	A	4569
	P	45996
Other Citrus	A	8169
	P	60996
Banana	A	39490
	P	518128
Mango	A	800
	P	4680
Guava	A	2870
	P	38005
Litchi	A	3968
	P	11920
Papaya	A	4882
	P	750558
Jackfruit	A	
	P	
Apple	A	
	P	
Pears	A	
	P	
Plum and Pears	A	
	P	

A = Area

P = Production

Source: Horticulture Statistics, National Horticulture Board, Ministry of Agriculture.
Govt. of India.

Table 10.3: District-wise Data of Fruit and Vegetable Units of Assam Since Inception to 31–03–98.

District	Fruit and Vegetable Based Units
Cachar	–
N.C. Hills	–
Sonitpur	2
Goalpara	2
Kamrup	14
North Lakhimpur	6
Karbi Anglong	2
Nagoan	17
Sibsagar	12
Dibrugarh	9
Kokrajhar	2
Dhubri	2
Darrang	–
Karimganj	1
Barpeta	13
Jorhat	–
Nalbari	1
Golaghat	1
Bongaigoan	3
Dhemaji	–
Hailakandi	–
Morigoan	–
Tinsukia	2

Source: Directorate of Industries, Assam.

Problems and Prospects of the Fruit and Vegetable Industry of Assam

Problems

The stunted growth of the fruit and vegetable processing industry of Assam is backed by various constraints. Some of the hurdles faced by the upcoming entrepreneurs of this industry are as follows:

1. The fruit and vegetable industry depends upon the local markets for the supply of raw materials. The quality, quantity and the price of the raw materials are constantly fluctuating and unpredictable in supply. Hence the production of the fruit and vegetable industry are confined to the seasonal availability of fruits and vegetables. The crops are also destroyed by the natural calamity that is floods which is a major problem faced by the agriculture of the state.

2. Another primary problem faced by this industry is finance. No adequate financial support is available from the government or from any private sector. It is also seen that the entrepreneurs suffer because of the hostile attitude of the banks and the other financial institution towards this industry.

3. Most of the food processing units of Assam are run on individual proprietary basis which is not a favourable condition to gain the advantage of economies of scale.

4. Lack of availability of packing materials is another important problem of this industry. Containers which are used for packing are not available within the state and are to be purchased from outside which increases the cost.

5. The units also face difficulty in labelling their products. Most of the units recourse to ordinary labelling as the rate of offset printing of labels being high, such cannot be afforded by the local units. Moreover the standard of offset printing of labels within the state is also not attractive. As a result, the products from outside the state are more attractively presented so far as packing and labelling is concerned and this attracts the local people to the products from outside the state.

6. Another major problem is the problem of warehousing of the products and the raw materials. The raw materials of this industry are mostly perishable and seasonal in nature and are to be preserved in cold storage but the entrepreneurs cannot afford the cold storage facility due to the lack of finance. Moreover there is also lack of space to store the finished products. Most of the units are started

by the entrepreneurs in their own sheds which are very small. Even the sheds allotted by the government in the industrial complexes are not sufficient for the units to store huge quantities of the finished products and raw materials.

7. There is also a dearth of skilled and semi skilled labour within the state.

8. There is no adequate research and development activities initiated by the local entrepreneurs to develop their products neither the government has shown much enthusiasm in this regard. No step has been taken for the upliftment of the quality of the products. Moreover, there is lack of FPO facility in the state (the nearest office being at Kolkata).

9. Another major constraint as to why the local products do not get a bright response from the consumers is due to the lack of advertisement. The entrepreneurs pay least interest to advertise their products and as a result people are less aware of the locally manufactured products.

10. The units are mostly ill organised and are started by the entrepreneurs in small scale at their own homes. Therefore the production of these units are not sufficient to cover a huge market.

11. Various taxes and duties both by the central and state government cause hurdle on the market expansion and creates problem in production especially for small units.

12. The local dealers and wholesalers do not give patronage to the local products because there is no suitable selling schemes for these products. The cost of production of the locally made products are quite high because of the increase in labour cost, hike in transportation rates and also due to the production processes.

13. Last but not the least, the psychological attitude of the local people that the local products are not adequate in comparison to the outside products creates hindrance in the development of the fruit and vegetable processing industry in the state.

Prospects

Though the fruit and vegetable processing industry of Assam are surrounded by many hurdles which creates problems in its development, yet it is however seen that there are also many prospects for the development of these industry in the state.

1. Abundance of raw materials within the state and the climatic condition suitable for setting up of fruit and vegetable industry.

2. The frequent change of food habits by the consumers increases production.

3. Large scale expansion of tourism and hotel industry in the region shows bright prospects of this industry.

4. The processing is restricted to only few fruits and vegetables. There is a vast potentiality to tap the under exploited fruits and vegetables in the state.

5. Moreover, the bright response secured by the entrepreneurs in the national and international fairs and exhibitions outside the state shows that if a little bit regard is paid by the entrepreneurs and the government towards the development of this industry it is certain that this industry will reach its highest peak and earn a fair amount of revenue to the state.

Strategies and Measures

1. There is a dearth of modern technology used by the industry. In order to increase productivity and meet consumer demand it is necessary to upgrade the technology.

2. To increase the horticulture productivity the government should provide improved seed at subsidize rate alone with other inputs like pesticides, fertilizers, agricultural implements. The government should also provide crop insurance scheme to provide relief to the farmers in case of loss by floods etc.

3. Periodic market survey should be taken by the entrepreneurs to find out the change of taste, price, competition etc.

4. Banks and financial institutions should liberalise the loan formalities and speed up the disposal of the loans.

5. Cold storage facilities should be provided by the government in various localities.

6. Multipurpose food testing centre and training is highly required. Moreover, research and development facilities should be highly developed.

7. The government should take initiative in the establishment of FPO (food processing order) office in the state vis-à-vis in the N.E. region.

8. The local entrepreneurs should visit the industrially developed states to get a first hand knowledge.

9. Both the central and the state government need to provide further reduction for taxes and duties particularly for packing materials since the whole of the N.E. region is industrially backward.

10. Efforts should be made both at government and private level for the exports of the products to other countries.

11. Fair and exhibitions should be arranged by the government to popularise the local products through display.

12. Preferential purchase order should also be strictly observed in case of any purchase by the government or state undertakings.

Conclusion

The fruit and vegetable processing industry of Assam is still in its developing stage though there are enough potentialities within the state for the development of this industry in terms of raw materials etc. yet the industry has to undergo a long process to establish itself in the national market. The various problems faced by this industry are to be collectively solved by the government and the private sector along with the semi government institutions to make the fruit and vegetable industry a landmark in the industrial scenario of Assam.

Chapter 11
Suggestions and Conclusion

The fruit and vegetable processing industry of Assam is yet to be started in an organised manner. The units are mostly existing as unorganised household ones entrusted with many problems. To overcome these problems of the processed fruit and vegetable industry a time has come to implement new policy incentives, initiatives and make investment from field to the processing unit and the consumer.

Therefore in order for the industry to grow and solve problems, the following strategy may be adopted both by the processor and the state government:

1. The present level of technology followed by the industry though not outdated are also not modern. In order to increase productivity and meet the stringent quality and consumer demands of the export as well as the local market it is necessary to upgrade the technology incorporating automation, better process control. Moreover, it is also required on the part of the government to encourage the entrepreneurs to start their units on an organised and large scale manner by giving various incentives and assistance.

2. Farmers should be motivated by creating awareness of the market value of their produce in order to improve their

productivity and quality. Moreover, in order to increase productivity government should provide improve seeds at subsidise rate. Furthermore, subsidies should also be provided for other inputs like pesticides, fertilizers, agricultural implements and equipment. Besides this, in order to provide relief to the farmers the government should draw a crop insurance scheme when the crops are destroyed by natural calamities like floods which are a common scenario in Assam.

3. The government should also take up supportive infrastructure scheme for post harvest handling and storage. In this context, mention can be made about the need of setting up cold storage facilities in the central place which is in an easy access to both the area of production and processing. Moreover, government should provide warehousing space in public warehouses in a nominal rate with first preference given to the fruit and vegetable industry.

4. The government should strengthen the backward linkage step by constituting cash incentive scheme for the best fruit and vegetable growers in order to ensure adequate supplies of raw material to the fruit and vegetable processing industries.

5. The government should take initiative to develop an organised market of the raw materials for the fruit and vegetable industry through the institution like NERAMAC so that it becomes easier for the growers to sell their produce and the processors to get standard raw materials for processing. Moreover, arrangements should be made by the government to render services of refrigerated vans while bringing the raw materials from the place of production to the market and the place of processing at concessional rate. Further, transportation facilities should be made easy for sending the finished products to the market throughout Assam and outside even during the rainy season when floods occur.

6. Periodic market survey should be conducted by the processor to find out changes of price, taste, competition etc.

7. The entrepreneurs of the fruit and vegetable processing industry should take up initiative for extensive advertisement of their products in order to create customer awareness of the products and capture the market.

8. Special care should be taken while packaging and labelling the products because outward presentation of the products have a telling effect on the minds of the customers.

9. Organised initiative should be undertaken for production of containers (either glass material or plastic material) for incessant and continuous flow of incidental requirements of successful fruit and vegetable based units.

10. Steps should be taken to modernise the labelling process of the fruit and vegetable based products by giving special attention by the Government for setting or providing assistance to the local unemployed youth for starting such units which will provide labelling through sophisticated techniques which will facilitate the local fruit and vegetable processing units.

11. Government should make arrangement for frequent visit of the local processors to the industrially developed state to get first-hand knowledge.

12. Both the central and the state government should make necessary arrangement for setting up of multipurpose food testing centre with research and development facilities.

13. Seminars and discussions should be organised by the government both at district and state level for a direct interaction between the growers and the processors as well as expert from the government and semi government institutions.

14. For the benefit of the entrepreneurs of the food processing industry of the North east the government should make arrangement for opening of the NE regional office of the FPO, CFTRI, Processed Food Export Promotion Council.

15. Steps should be taken by the government institutions like APEDA to register more and more units under ISO 9000 for exporting the products outside India. Moreover, national level seminars and workshops should be

organised where reputed exporters and government officials of the particular industry should be invited to give latest information regarding exports to the local entrepreneurs. Further first-hand information should be provided to the entrepreneurs regarding the norms to be followed during the processing of the products meant for exports by the government institutions like APEDA.

16. In order to export the products and capture the International market the entrepreneurs should be given training on the International quality standards and about the hygienic way of processing the fruits and vegetables for the acceptance of those by the advanced countries. Moreover, they should be given training on the technical side of such processing, for example, uniform slicing of the products, using of correct apparatus and utensils required for processing and expert know how of processing the raw materials into finished products.

17. The central government should provide reduction of taxes and duties particularly for packaging material for the North east region since the region is industrially backward.

18. Bank and financial institutions should liberalise the loan formalities and provide speedy disposal of loan to the local entrepreneurs/processors.

19. It is observed that the entrepreneurs are not fully aware of the various schemes provided by the central and the state government, therefore efforts should be made for the proper dissemination of the information regarding such schemes by the government.

20. The retailers should be given an opportunity of credit system as well as privileges of sending back unsold materials which loose quality owing to long gap between production and sale by the entrepreneurs.

21. Efforts should be made on the governmental level to provide training by the experts on the correct blending of chemicals and other ingredients required for taste, durability and preservation of the product as the entrepreneurs have not earned expert knowledge in correct blending of such.

22. The Government should take up measures to control, flood in the state and try to restore the transportation of the raw materials and the finished products even during the rainy season.

Conclusion

The fruit and vegetable processing industry plays a significant role in the socio-economic development particularly in the rural areas of Assam. Despite abundant natural resources as raw materials for this industry, it is still in a very poor condition. This is because of the casual operation, unscientific and uncoordinated approach from the field to the processor and the consumer. So there is a need to make a joint effort on the part of the entrepreneurs and the Government to over come these constraints for the future growth of the industry.

Bibliography

Books

Aggrawal Ravikiran (Vice Chairman and MD Tasty Bite Eatables Ltd, Mumbai); Special needs of problems in the marketing of processed foods; Central food technological research institute, Mysore; 1989.

Banerjee Mrityunjoy; Essentials of modern marketing; Oxford and IBH publishing Co. Pvt. Ltd; 1988.

Borthakur Dhirendra Nath; Agriculture in the North Eastern region with special reference to hill agriculture; BeeCee Prakashan, Guwahati; 1992.

Crimp Margeret; The marketing research process; Prentice Hall International; 1991.

Desai B.M., Gupta V.K., Nambodi N.K; Food processing industry: Development and financial performance; Oxford and IBH publishing Co. Pvt. Ltd., New Delhi, Kolkata, Mumbai; 1991.

Dr. Shukla S.P., Dr. Aggarwal A.K.; Agriculture of N.E. region; National Publishing House; 1986.

Dala V.B.; Pretreatment, packaging, storage and transportation of fresh foods and vegetables (Trends in food science and technology) Consultant, formerly scientist at CFTRI, Mysore, Association of food scientist and technologist (India) CFTRI, Mysore; 1989.

Desai B.M.; Food processing industries, Development and financial performance, New Delhi, Oxford; 1991.

Das M.M.; Peasant Agriculture in Assam; Inter-India Publications, New Delhi; 1984.

Fisk Geogre; Marketing system: An introductory analysis; Harper and Row Publisher; 1967.

Goswami P.C.; Agriculture of Assam; Assam Institute of Development Studies; 1989.

Ghosh S.P.; Horticulture in North-Eastern India; Associated Publishing Company, New Delhi; 1985.

Jones Stephen; Food markets in Developing Countries, What we know? Oxford International Development Centre; 1996.

Kotler Philip, Marketing Management Analysis, Planning, Implementation and Control; Prentice-Hall of India Pvt. Limited; 1992.

Krishnamurty G.V. New product from food and vegetable processing waste; CFTRI, Mysore; 1980.

Mamoria C.B, Joshi R.L; Principles and practice of marketing in India; Kitab Mahal; 1997.

Malhotra Vikram; Food processing industries in India, Investment opportunities; Global; 1993.

Mamoria C.B, Mamoria Satish; Marketing Management; Kitab Mahal; 1997.

Marquadt. A. Raynand, Makens C. James, Roe G. Robert; Retail Management, Satisfaction of consumer needs; The Dryden Press; 1983.

Poulty V.H and Mukly M.J; Food processing; Oxford and IBH Publishing Co. Pvt. Ltd., New Delhi, Kolkata, Mumbai; 1993.

Ramasyamy V.S.; Namakumari, S.; Marketing Management, Planning, Implementation and Control; Macmillan India limited; 1994.

Rajajee, M.S., Rasheed, I.A., Narasimhan, S.; Export marketing and management; S. Chand and Company, New Delhi; 1982.

Sherlekhar S.A.; Marketing Management; Himalaya Publishing House; 1991.

Sarma Keshav; Marketing Management of Horticulture Produce; Deep and Deep Publications; 1991.

Stanly E Richard, Promotion, Advertising, Publicity, Personal selling, Sales Promotion; Prentice Hall International; 1982.

Sontakki C.N.; Marketing Management, Kalyani Publisher; 2000.

SIRI Board of Consultant of Engineers; Food processing industry, Small Industry Research Institute, Delhi; 1994.

Srivastava V.K., Patel N.T.; Managing food processing industry in India; Oxford and IBH publishing Co. Pvt. Ltd.; 1994.

SPB, Board of Consultant and Engineers, Handbook of export oriented food processing projects, Publication Division SPB Consultants and Engineers Pvt. Ltd., Delhi; 1994.

Talukdar K.C., Bhowmick B.C. (Eds); Marketing of Perishable Products; B.R. Publishing Corporation, Delhi; 1993.

Veerajee P.; Food Packaging: New Trends; CFTRI; 1989.

Articles

Bhatnagar P.S.; Food processing industries; A sunrise sector; Employment news; March, 2000.

Choudhury Monalisa and Dr. Barua Nayan; Strategy for Development of Fruit and Vegetable based Industry in Assam; Journal of the North Eastern Council; Vol. 19, No. 3, July–Sept., 1999.

Choudhury Monalisa and Dr. Barua Nayan; Food processing industry in the North East: Its Prospects and Problems; Journal of Commerce, Gauhati University, Vol. 7, 1999.

Food processing industry: Promising trends, Kisan World 22(7); July, 1995.

Goshal M.K; Food processing: Thrust area for export and foreign collaboration, Yojana 34(18); 1 October, 1990.

Kapur S.L.; Good scope for food processing industries; Economic Times (Supplement) 4 February, 1992.

Mukherjee Dhurjati; Food processing industry Can India Play a leading role. Khadi Gromodyog 3(11); August, 1990.

Nand Kumar M. and Ray Shyamal; Food processing: Need to go beyond adhocism; Economic times; 30 November, 1992.

Parthasarthy Ashok; Food processing industry, Futuristic view, SEDME 25(1); March 1998.

Ram Bhadra; Agro Foods; Is there a export potential? Economic times; 8 December, 1990.

Ramaswamy Vasant Kumar and Katula Rajni; Processed foods: A system approach to policy and development; Indian Journal of Agriculture Economics 47(3); July–Sept., 1992.

Shah Narendra; Opportunities and Challenges in agro based industries: Food processing; Journal of Rural Development 17(3); July–September, 1998.

Singh H.P. Potentialities of food processing industry in India; Economic affairs 34(3); July–Sept., 1989.

Sud Surinder; Indian likely to be World's largest food factory; Indian perspective 11(3); March, 1998.

Vishwanathan K.V. and Satyasai K.J.S.; Fruit and Vegetables: Production trends and role of linkages; Indian Journal of Agriculture Economics 52(3); July–Sept., 1997.

Vyas D.M. and Patel J.C.; Rural development through Food processing; Yojana 35(7); 30 April, 1991.

Journals

Beverage and food world; Dairy management Consultant.

Indian food industry; Publication of association of food scientist and technologist (India).

India spices; Spices Board of India; Ministry of Commerce, Govt of India.

Indian food Packer; All India food processors association.

Journal of the North East Council.

Processed food industry, Publisher and Printer, Salem Akher Taqvi.

Small Enterprises Development and Management Extension (SEDME). Vol. 25, No. 1 March, 1998.

Government Report and Documents

Agriculture and Processed food Products Export Development Authority (APEDA) Quarterly Information line.

Abstract of Guidelines for Manufacturing licence; Published by Government of India, Ministry of food processing industries, Office of the Deputy Director (Food and vegetable processing) Eastern Region, Kolkata.

Basic Statistics of North Eastern Region; North Eastern Council, Shillong; 1995.

Basic Statistics of Assam; Directorate of Economics and Statistics, Assam; 1997.

Getting in touch with Assam Small Industries Development Corporation (ASIDC); Printed and published by ASIDC, Bamunimaidan, Guwahati.

Industrial Policy of Assam; Published by Assam Industrial Development Corporation Ltd on behalf of Industries Department, Government of Assam, 1997.

Indian Institute of Entrepreneurship (IIE) At a Glance; Published by IIE, Guwahati.

Information on the preservation of various fruit and vegetable processed products; Published by the Agriculture Department, Government of Assam.

National Small Industries Corporation Limited (NSIC) Corporate Profile; Published by NSIC, Government of India Enterprise, New Delhi.

North Eastern Industrial and Technical Consultancy Organisation Ltd (NEITCO); Published by NEITCO, Udyog Bikash Bhawan, G.S Road, Bhangagarh.

Operation Guidelines; National Horticulture Board, Ministry of Agriculture, Government of India, August 2000.

Schemes of Assistance of the Small Industries Development Bank of India.

Schemes at a Glance, North Eastern Development Finance Corporation Ltd. The Development Bank of and for the North East.

Index

www.ingramcontent.com/pod-product-compliance
Lightning Source LLC
Chambersburg PA
CBHW070709190326
41458CB00004B/912